T0133401

Applied Electron Microscopy
Angewandte Elektronenmikroskopie

Band 8

Applied Electron Microscopy
Angewandte Elektronenmikroskopie

Band 8

Prof. Dr. Josef Zweck (Hrsg.)

Jürgen Gründmayer

TEM Investigations on Magnetic and
Structural Phase Transitions at Low Temperatures

Logos Verlag Berlin

λογος

Applied Electron Microscopy
Angewandte Elektronenmikroskopie

Herausgeber:
Prof. Dr. Josef Zweck
Institut für Experimentelle und Angewandte Physik
Universität Regensburg
93040 Regensburg
Germany
Email: josef.zweck@physik.uni-regensburg.de

Bibliografische Information der Deutschen Nationalbibliothek

Die Deutsche Nationalbibliothek verzeichnet diese Publikation in der Deutschen Nationalbibliografie; detaillierte bibliografische Daten sind im Internet über http://dnb.d-nb.de abrufbar.

© Copyright Logos Verlag Berlin GmbH 2009

Alle Rechte vorbehalten.

ISBN 978-3-8325-2270-4
ISSN 1860-0034

Logos Verlag Berlin GmbH
Comeniushof, Gubener Str. 47,
10243 Berlin

Tel.:   +49 (0)30 / 42 85 10 90
Fax: +49 (0)30 / 42 85 10 92
http://www.logos-verlag.de

Preface to the series "Applied Electron Microscopy /
Angewandte Elektronenmikroskopie"

The laboratory for electron microscopy at the physics faculty (University of Regensburg, Germany) works mostly on problems in solid state physics using various electron microscopic methods. Our starting point is that an electron microscope measures twodimensionally the interaction between electrons and the specimen. The result of this measurement is the usual "image".

As is well known and generally accepted, a purely qualitative and seemingly "straight forward" interpretation of such "images" is at least dangerous and in most cases misleading. However, modern electron microscopy offers manyfold possibilities to extract physical properties from the images. This is possible now due to the well established electron optical theory of contrast formation and the wealth of different techniques and detectors available which enable the experimentator to precisely investigate the specific interactions between electrons and the specimen which are of interest for a specific problem.

To achieve this goal, it is frequently necessary to use theoretical and/or purely academic demonstrations of the feasibility of certain effects to develop methods and procedures which are suited to solve a specific problem. In the past many theses were written which produced new results both for electron microscopy and solid state physics. Following a suggestion of my students and with friendly support of the Logos publishing company I decided to publish these theses as a series "Applied Electron Microscopy / Angewandte Elektronenmikroskopie" to prevent that the theses which have been worked upon with diligence and hard work end up to soon in some library storage rooms. They could be of value elsewhere, if only someone knew that they exist.

I would be pleased if we could attract contributions also from other institutes, laboratories, facilities etc. which may be the starting point of a series of books where one can find fast many recent theses on electron microscopy. Although the title of this series may suggest otherwise, we welcome all theses which treat state-of-the-art electron microscopy, no matter whether this is related to theory, instrumentation, methodology or the solution of a specific problem. It goes without saying that both theses in German and in English will be accepted.

Regensburg, December 2004

Prof. Dr. Josef Zweck

TEM Investigations on Magnetic and Structural Phase Transitions at Low Temperatures

Dissertation zur Erlangung des Doktorgrades der Naturwissenschaften
(Dr. rer. nat.) der Fakultät Physik der Universität Regensburg

vorgelegt von

Jürgen Gründmayer

aus Eggenfelden

durchgeführt am Institut für
Experimentelle und Angewandte Physik
der Universität Regensburg
unter Anleitung von
Prof. Dr. J. Zweck

Juni 2009

Promotionsgesuch eingereicht am: 03.06.2009

Tag der mündlichen Prüfung: 22.07.2009

Die Arbeit wurde angeleitet von Prof. Dr. Josef Zweck

Prüfungsausschuss: Prof. Dr. John Schliemann (Vorsitzender)
Prof. Dr. Josef Zweck   (1. Gutachter)
Prof. Dr. Christoph Strunk (2. Gutachter)
Prof. Dr. Sergey Ganichev (Prüfer)

PHANTASIE
IST WICHTIGER ALS WISSEN,
DENN WISSEN IST BEGRENZT.

Albert Einstein, * 14. März 1879, † 18. April 1955

Contents

1 Introduction and Motivation

Phase transitions are a physical phenomenon known by mankind from its very beginning 200,000 years ago with the transition of H_2O from solid ice to liquid water to gaseous steam. These three manifestations are called *states of matter*, of which the fourth one – discovered in the modern times – is plasma. The reason for a phase transition is a change in the environmental conditions like temperature and / or pressure and is characterized by a phase transition point at which the material abruptly changes its properties. But the transitions between the so called states of matter are not the only phase transitions, there are a lot more, and amongst them we find the transition between the ferromagentic and paramagnetic phase of magnetic materials as well as the crystallographic martensitic phase transition. These two phase transitions are associated with microscopic structural changes and this makes them very interesting candidates for detailed investigations with a transmission electron microscope (TEM).

The starting point of our microscopic investigations of phase transitions will be the diluted magnetic semiconductor (DMS) (Ga,Mn)As which is one of the most prominent materials of the Sonderforschungsbereich 689 "Spin Phenomena in Reduced Dimensions" in whose context the main part of this work was carried out.

Diluted magnetic semiconductors are the key component for the emerging technology of spintronic (a neologism meaning "spin transport electronics") devices, which combine the properties of classical semiconductor devices which make use of solely the charge of an electron by using although its spin and its associated magnetic moment, adding one degree of freedom for information processing. All the future devices rely heavily on the knowledge of the micromagnetic configurations, switching characteristics and interactions between differently shaped microstructures made of DMSs like (Ga,Mn)As and their dependence on temperature.

It was clear from the start that (Ga,Mn)As would be no easy matter for magnetic TEM investigations and that lots of new techniques will have to be developed and tested on their practicability. Although the Curie temperature at which (Ga,Mn)As becomes ferromagnetic could be pushed up to 160 K and even more [Sad06b], the typical values of T_C are more in the range of 50 K - 90 K, especially for our samples. This and the fact that the already weak magnetic moment of (Ga,Mn)As is even weaker close to the transition temperature made it inevitable to use a new liquid helium cooled sample holder. The verification of the compatibility of this holder with advanced magnetic imaging techniques like electron holography will be one subject of this work.

A second topic which required a lot of spadework was the preparation of electron transparent TEM specimens out of the (Ga,Mn)As samples we received from other groups as the standard procedures implicate an unknown amount of thermal stress on the specimen, which influences the physical properties of (Ga,Mn)As heavily. This made it necessary to develop specially adapted methods for thinning those samples.

It was a natural step to apply the experience we gathered during our work on this extremely delicate material to other physical systems. The first one being another diluted ferromagnet used within the Sonderforschungsbereich with a Curie temperature well below room temperature, which is an alloy of palladium with a few percent of iron. This material is a promising candidate for contacting carbon nano tubes for spin coherent carrier injection. Contrariwise to our first matter of research, PdFe is nearly ideal for the investigation of its magnetic phase transition with a TEM. Its magnetic moment is many times the one of (Ga,Mn)As, so thin layers of 50 nm deliver a sufficient magnetic signal. Small changes in the iron concentration vary its transition temperature by several tens of Kelvin, making it possible to use a liquid nitrogen cooled sample holder that is much easier to handle than the helium cooled one and thus allowing prolonged experiments when using iron concentrations greater than 3 %. A special sample preparation is not necessary for PdFe, as it can be directly grown on top of commercially available Si_3N_4-membranes.

The third subject was processed within a collaboration with a group of the University of Duisburg-Essen. Their aim is to prove that superconductivity goes along with a certain crystallographic phase transition. The investigation of this transition which was expected to produces a fine structure at a length scale of several nanometers at temperatures below 100 K fitted right into the context of this work while enabling us to extend the sort of the examined phase transitions to a pure structural one.

The following chapter generally introduces the principle of phase transitions and is followed by a theoretical overview on the magnetic phenomena appearing in solids. Chapters 4 and 5 explain the functionality of the instruments used in this work, namely our transmission electron microscope FEI Tecnai F30 Regensburg Special and the sample holders with accessories. After those general parts, each material system is introduced in detail and the results are presented in one chapter for each material. As PdFe with its crystallographic phase transition is different from the main parts concerning magnetic phenomena, its chapter includes a basic introduction to the martensitic transition. Finally, the findings are summarized and outlook on future TEM experiments related to phase transitions is given.

2 Introduction to Phase Transitions

The following section will provide a quick overview of the most important principles and terms related to *phase transitions*; more detailed information can be found in [Geb80]. Generally said, a phase transition is the transition from one state of matter to another. This transition is characterized by a certain point at which a physical property undergoes an abrupt change. At this point, the *free energy* of a system is non-analytic for some choice of order parameter (e.g. the magnetization) which is usually a result of the interaction of a large number of particles in the system.

Classification Schemes

As phase transitions occur in many different forms which have to be regarded seperately, they are grouped into different classes. A first thermodynamical classification was proposed by Ehrenfest and is based on the degree of non-analycity. Phase transitions of first order show a discontinuity in the first derivative of the free energy, second-order phase transitions are continuous in the first derivative but discontinuous in the second. According to this scheme, e.g. the solid - liquid - gas transitions are of first order while the ferromagnetic phase transition is of second order.

In the cases when the derivative of the free energy diverges, the Ehrenfest scheme is no longer sufficient. The modern classification scheme still uses the same notations, but different definitions.

Phase transitions are now called first-order when they involve latent heat. During the transition, the system either absorbs or releases a certain amount of energy but the temperature of the system does not change. The fact that the energy transfer between the system and its environment takes some time leads to mixed-phase regimes, in which parts of the system have completed the phase transitions while others have not. Second-order phase transitions, also called continuous phase transitions, do not involve latent heat. The *Landau-theory* provides a phenomenological description of those transitions.

Symmetry Breaking

Many phase transitions go together with the phenomenon of symmetry breaking which happens when a system changes from a disordered to an ordered state. The paramagnetic to ferromagnetic phase transition (explained in more detail in the following chapter) is one example. After switching to the ferromagnetic state, the system shows a preferred direction, which is the direction of the *magnetization* $\vec{M}$. The paramagnetic system has the same crystal symmetry as the ferromagnetic system, but its spins are rotational invariant as every single spin points to every direction over the corse of time. The ferromagnet is only invariant for rotations parallel to $\vec{M}$ (see fig. 2.1). The important

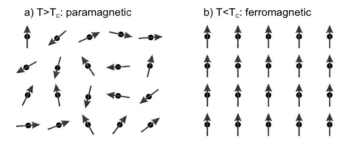

Figure 2.1: *Transition from paramagnetic (a) to ferromagnetic order (b) occuring at the transition temperature T_C. The arrows indicate the orientation of the single spins.*

point is that changes in symmetry can not happen gradually, as the given symmetry either exists or not. This means that those phase transitions have an exact border between the ordered and the disordered state – phenomena appearing with phase transitions not related to symmetry breaking like the *critical point* of gases and liquids at which both phases coexist cannot happen here.

Order Parameter

The order parameter is a physical quantity that changes significantly between two phases, usually it is 0 in one phase and non-zero in the other. This parameter characterizes the rise of order when the phase transition occurs. Its *susceptibility*[1] usually diverges at the transition point.

In the case of symmetry breaking, one or more additional variables have to be introduced to describe the state of a system. The ferromagnetic transition is an example for this, because the magnetization is a value that suddenly appears below the phase transition temperature.

---

[1]The *susceptibility* is a physical quantity characterizing the magnetizability of a material by an external magnetic field.

3 Magnetism in Solids

This chapter gives an introduction to the theoretical concepts of magnetism in solids which are necessary to understand the experiments of chapters 6 and 7.

The first part introduces the ferromagnetic order and outlines the according phase transition. From there on, the temperature is assumed to be far below any relevant transition temperature, and thus every specimen is assumed to be ferromagnetic.

Next, the different energies which determine the magnetic properties of a sample are described. The result of their interaction is the formation of a certain magnetization configuration in a way that the sum of the involved energies is minimized. After that, the calculation of the contribution of the shape anisotropy is outlined. The following rigid-vortex-model is a theory describing the magnetization changes of magnetic discs in the vortex configuration as we will observe in section 7.6. Finally, the transformation of the magnetic energy landscape at temperatures slightly below T_C is discussed.

More detailed information on the magnetism of small particles can be found in [Höl04] and [Sch03].

The absolute value of the magnetization $\vec{M}(\vec{r})$ can be considered equal to the saturation magnetization M_S [Bro62]. Thus, it is sufficient to use the vector $\vec{m}(\vec{r}) = \vec{M}(\vec{r})/M_S$ specifying the direction of the magnetization $\vec{M}(\vec{r})$ at the location $\vec{r}$ with the boundary condition $(\vec{m}^2(\vec{r}) = 1)$. In the following, $\vec{m}(\vec{r})$ will be called magnetization, too.

3.1 Ferromagnetic Order and Phase Transition

The characteristic property of a ferromagnet is that it has a spontaneous magnetic moment which exists without any external magnetic field. This magnetic moment is called *saturation moment* and is present if there is an inner interaction (called *exchange field* which makes the magnetic moments align parallel. The opponent to this ordering interaction is the thermal movement which destroys the order at higher temperatures. The exchange field can be considered as equal to a magnetic field $\vec{B}_E$. The *molecular field approximation* assumes that

$$\vec{B}_E = \lambda \mu_0 \vec{M} \tag{3.1}$$

with λ as a temperature dependent constant. According to this formula, each spin is influenced by the mean magnetization caused by all the other spins.

The *Curie temperature* T_C is the specific temperature above which the spontaneous magnetization vanishes, and thus it separates the ferromagnetic from the paramagnetic phase. For the latter case, an external field B_{ext} produces a finite magnetization which results in an exchange field B_E. If χ_p denominates the paramagnetic susceptibility, we get

$$\mu_0 M = \chi_p(B_{ext} + B_E). \tag{3.2}$$

This equation is only valid in the paramagnetic case where the share of the aligned magnetic moments is small. With the Curie-constant C^1 and the Curie-law $\chi_p = C/T$, we get $\mu_0 M T = C(B_{ext} + \lambda\mu_0 M)$ and can now calculate the susceptibility

$$\chi = \frac{\mu_0 M}{B_{ext}} = \frac{M}{H} = \frac{C}{T - T_C} \tag{3.3}$$

with $T_C = \lambda C$. Equation 3.3 is called the *Curie-Weiss-law* and is a good description for the susceptibility above the Curie temperature. The susceptibility has a singularity at $T = T_C$. Below that temperature, a spontaneous magnetization exists, because it is possible that with an infinite χ and a vanishing B_E a finite M exists.

More detailed calculations (see [Kit02]) deliver

$$\chi \propto -\frac{1}{(T - T_C)^{1.33}} \tag{3.4}$$

for temperatures close to T_C.

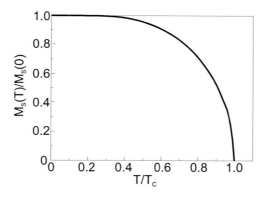

Figure 3.1: *Saturation magnetization as a function of the temperature calculated from equation 3.5.*

The molecular field approximation can also be applied to temperatures below T_C. In this case, instead of the simplified Curie-law, the *Brillouin expression* has to be used, which is according to [Kit02] given by $M = N\mu\tanh(\mu\lambda M/k_B T)$ for N particles with

---

[1] $C = \mu_0 n \frac{\mu^2}{3k_B}$ with the particle density n, the atomic magnetic moment μ, the vacuum permeability μ_0 and the Boltzmann-constant k_B.

spin $1/2$. Ignoring the external field and substituting B by the molecular field $B_E = \lambda M$, we get

$$M = N\mu \tanh(\mu \lambda M / k_B T).\tag{3.5}$$

A graphical solution for this equation can be found in [Kit02] and shall not be given here. The resulting curve for the normalized magnetization is shown in fig. 3.1 and characterizes a second order phase transition.

3.2 Contributions to the Free Energy

The free energy of a ferromagnet is composed of different energy contributions:

$$E_F = E_{ex} + E_Z + E_{ani} + E_D \tag{3.6}$$

with the exchange energy E_{ex}, the Zeeman energy E_Z, the sum of the anisotropy energies E_{ani} and the stray field energy E_D.

3.2.1 Exchange Energy

The phenomenon of ferromagnetism is based on the so called *exchange interaction*. In its most simple form, we regard neighboring spins whose parallel orientation has the lowest energy and thus we get as a result the magnetization $\vec{M}$ inside our sample. The exchange energy E_{ex} in a system with spins of the same absolute value S is given by Heisenberg as

$$E_{ex} = -\sum_{i>j} J_{ij}\vec{S}_i\vec{S}_j = -\sum_{i>j} J_{ij}S^2\cos\phi_{ij} \tag{3.7}$$

with the exchange integral J_{ij} and the angle ϕ_{ij} between the spins S_i and S_j. The coupling is ferromagnetic for $J_{ij} > 0$ and antiferromagnetic for $J_{ij} < 0$.

The Heisenberg-model is only valid for localized spins and already incorrect for 3d-transition metals like iron or nickel. Therefore, the band model has to be used which delivers an equivalent form of the exchange energy:

$$E_{ex} = A \int_V (\operatorname{grad} \vec{m}(\vec{r}))^2 \, \mathrm{d}V \tag{3.8}$$

with the material specific exchange constant A arising from the exchange integral J_{ij} and the sample volume V.

RKKY Interaction

In the section above, we only considered the direct exchange between magnetic moments. This is only a valid description when the moments are close together. In situations where the mean distance of the magnetic moments is large enough that the exchange integral practically vanishes, this model is no longer correct. This is the case for the two magnetic materials investigated in this work, (Ga,Mn)As and PdFe, where the magnetic dopants have concentrations around 5% and thus they are too far away from each other to interact directly.

The exchange mechanism applicable in this case was discovered by Ruderman and Kittel in 1954 for the atomic core spins and the conduction band electrons and later generalized by Kasuya and Yosida in 1956 and 1957. The first letters of these four people give the interaction mechanism its name: RKKY.

In the case of the RKKY interaction, the spin of a free carrier is aligned by a localized magnetic moment and while moving through the crystal lattice, this free carrier polarizes other local magnetic moments. So the interaction between the localized magnetic moments is carrier mediated and we call this an *indirect exchange mechanism*.

The basis of the RKKY exchange has an already familiar form, given by the effective Hamiltonian

$$\mathcal{H}^{RKKY} = -\sum_{i<j} J_{ij}^{RKKY} \, \vec{S}_i \vec{S}_j \, . \tag{3.9}$$

Hereby, the electrons characterized by their Bloch wave functions mediate the interaction between the spins $\vec{S}_i$ und $\vec{S}_j$. In order to determine the coupling constant J_{ij}^{RKKY} between these spins, second order perturbation theory is necessary. The derivation not given here (see [Rud54], [Kas56] and [Yos57] for details) delivers

$$J_{ij}^{RKKY} = \frac{g^2 k_F^6}{E_F} \cdot \frac{\hbar^2 V^2}{N^2 (2\pi)^3} \cdot \frac{[\sin(2k_F r_{ij}) - 2k_F r_{ij} \cos(2k_F r_{ij})]}{(2k_F r_{ij})^4} \tag{3.10}$$

with the electron g-factor, the reciprocal spin density V/N, the Fermi-energy E_F, the Fermi-wave-vector k_F and the distance between two magnetic moments r_{ij}. Fig. 3.2

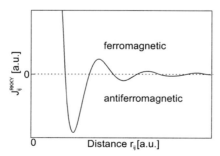

Figure 3.2: *The RKKY exchange integral versus the distance of the magnetic moments. One can see the characteristic oscillations and the $1/r^3$ decay.*

schematically shows the RKKY exchange integral with its characteristic oscillations between ferromagnetic and antiferromagnetic coupling and the $1/r^3$ decay.

As the Fermi wave vector $k_F = (3\pi^2 n_e)^{4/3}$ as well as the Fermi energy $E_F = \frac{\hbar^2}{2m^*}(3\pi^2 n_e)^{1/3}$ contains the electron density n_e, we see that $J_{ij}^{RKKY} \sim n_e^{4/3}$. The carrier density thus has a crucial influence on the indirect exchange interaction and hence on the magnetic properties.

3.2.2 Zeeman Energy

An external magnetic field $\vec{H}_{ext}$ applies a torque on the magnetization $\vec{M}$ which tries to align the magnetization along the external field (see fig. 3.3).

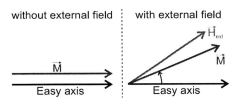

Figure 3.3: *The magnetization $\vec{M}$ is deviated from the easy axis by the external magnetic field $\vec{H}_{ext}$.*

The resulting energy

$$E_Z = -\mu_0 \cdot M_S \cdot \int_V \vec{m} \cdot \vec{H}_{ext}\, \mathrm{d}V \tag{3.11}$$

with the magnetic field constant μ_0 is called *Zeeman energy*. It is minimal for $\vec{H}_{ext} \parallel \vec{M}$.

3.2.3 Anisotropy Energies

The anisotropy energy depends on the direction of $\vec{M}$ in respect to different axis of the magnetic structure. When the anisotropy energy has its minimum, $\vec{M}$ is parallel to the so called *easy axis*. The direction which has a maximum anisotropy energy is consequently called *hard axis*. If only one prominent axis exists, this situation is called *uniaxial anisotropy*. For polycrystalline materials, the physical shape of the sample plays a dominant role in relation to the other anisotropies.

The impact of the crystallographic lattice on the energy of a ferromagnet is called *crystal anisotropy*. Because of the interaction between the electron spins and the magnetic orbital moments, certain orientations of the magnetization in respect to the crystal lattice are energetically favorable and thus the magnetization tries to align itself accordingly. For cubic crystals as an example we get

$$E_{cub} = V\left(K_1(m_x^2 m_y^2 + m_x^2 m_z^2 + m_y^2 m_z^2) + K_2(m_x^2 m_y^2 m_z^2)\right) \tag{3.12}$$

where the m_i denominate the components of the magnetization referring to the crystal axis. K_1 and K_2 are material constants and correspond to an energy density. The signs of this constants determine the easy or hard axis for the different orientations in the crystal. The uniaxial anisotropy as mentioned before has only one prominent orientation which can be either a hard or an easy axis. This can be described by

$$E_{uni} = V \cdot (K_U \sin^2 \phi) \tag{3.13}$$

with the uniaxial anisotropy constant K_U and the angle ϕ between the magnetization and the prominent axis. K_U is positive for the easy axis and negative for the hard axis.

For the PdFe used in this work, the crystal anisotropy can be widely neglected as the crystallites are randomly oriented and distributed which leads to a nullification. But one cannot rule out that the sample gets an uniaxial anisotropy imprinted during the growth process in the MBE by an unintended external magnetic field.

3.2.4 Stray field Energy

The spatial distribution of magnetic structures in the sample can lead to stray fields outside the sample or demagnetizing fields within the sample itself ($\vec{H}_D$) which are counteracting the magnetiztion $\vec{M}$. Maxwell tells us that $\nabla\vec{B} = 0$, and with $\vec{B} = \mu_0(\vec{H} + \vec{M})$ we get

$$\nabla\vec{H}_D = -\nabla\vec{M} . \tag{3.14}$$

This shows us that the sinks and sources of the magnetization act as poles for the stray field. This can be the case when e.g. the magnetization has a component perpendicular to the boundary at the rim. The resulting stray field can be calculated and referring to [Bro62], we get for the *stray field energy*

$$E_d = \underbrace{\frac{1}{2}\mu_0 \int_{Space} \vec{H}_D^2(\vec{r})\,\mathrm{d}V}_{1} = \underbrace{-\frac{1}{2}\mu_0 M_S \int_V \vec{m} \cdot \vec{H}_D(\vec{r})\,\mathrm{d}V}_{2} . \tag{3.15}$$

Integral 1 is hereby integrated over the whole space, integral 2 only over the sample's volume. From the left side, we can easily see that each contrubution to the stray field increases the energy.

3.3 Magnetic Domains

The magnetization of a ferromagnetic material is aligning itself spontaneously because of the exchange interaction. If all spins are pointing in the same direction, magnetic poles are forming at the rim of a particle which generate a stray field (see fig. 3.4 a). This increases the total energy of the magnetic particle. By splitting the magnetization into several areas with different magnetization directions, called *magnetic domains*, the stray field energy can be reduced. A domain configuration with a closed flux (see fig. 3.4 c) has a minimal stray field energy. But this configuration has an increased exchange energy at the domain boundaries where the spins are no longer aligned parallel.

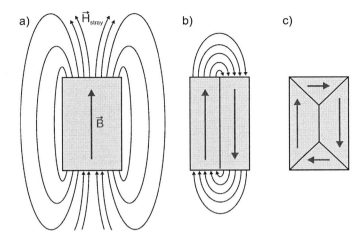

Figure 3.4: *Stray fields for different domain configurations. The stray field is becomes smaller from a) to c), c) itself has no stray field at all because of its closed magnetic flux. The upper and lower domains of c) are called closure domains.*

This change of the spin orientation is not abruptly, there is a transition area in which the direction of the spins changes continuously. This area is called a *domain wall*. Depending on the probe geometry, there are two possible ways how the rotation from one to the other direction can take place (see fig. 3.5): In thin films, the magnetization is rotating within the sample plane, and the domain boundary is called *Néel-wall* in that case. With thicker films, it is energetically more favorable to rotate the magnetization in the wall out of the sample plane. This is called a Bloch-wall.

Between these two wall types, there is room for the so called *crosstie-walls* (see fig. 3.6) which are a combination of Bloch- and Neél-walls. The domain wall thus consists of a series of vortices (compare section 3.3.1) and anti-vortices. These walls can be found in soft magnetic films with thicknesses of around several exchange lengths[2].

When speaking of a domain configuration, we mean that there are areas with uniform magnetization which are separated from each other by distinct walls. The width of a domain wall is typically well below $1\,\mu$m, and having particles of this dimension, the term domain configuration is no longer suggestive. As the magnetization is then rotating over the whole sample, magnetization configuration is the better terminology.

---

[2]The exchange length characterizes the range of the magnetic interaction between the single micromagnetic moments. It depends mainly on the exchange stiffness and the anisotropy constant and is usually in the range of several to several hundred nanometers.

a)

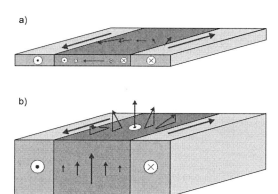

b)

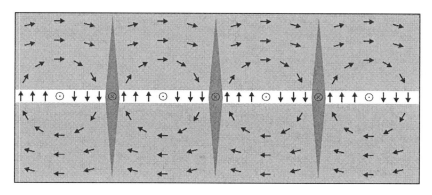

Figure 3.5: *Thin ferromagnetic layers can show two different types of domain walls (after [Heu05]): Very thin films have so called Néel-walls where the magnetization is rotating within the sample plane (a), thicker films have Bloch-walls with a rotation pointing out of the sample.*

Figure 3.6: *Micromagnetic configuration of a crosstie-wall. The different grayscales indicate a TEM image in Lorentz mode (see chapter 4). The magnetization lays in the sample plane, except at the intersections and the middle of the intersections where we find Bloch lines.*

The competition between anisotropy energy and stray field energy determines the magnetization configuration of a magnetic film without an external magnetic field. The reduction of the stray field by forming domains is always correlated to a rising anisotropy energy at the walls.

When an external magnetic field is applied, the magnetization is aligning parallel to this field and by that reducing the Zeeman energy and consequently the total energy. The net magnetization of the particle is then pointing in the direction of the external field. But as this reorientation increases the stray field energy, an equilibrium situation is achieved again with a minimal sum of energies.

When the external field is changing, the magnetization is adapting in a way that a global or local minimum is achieved and stays in that configuration. If two minima are only separated by a small energy barrier, thermal effects can make this configuration unstable and thus force the magnetic configuration to switch between different states. Details and experimental results on this can be found in chapter 7.7.

3.3.1 The Vortex State

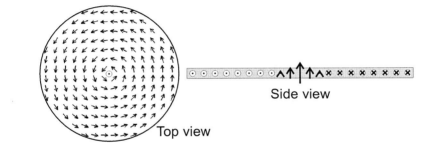

Figure 3.7: *Vortex magnetization configuration. The magnetization is rotating in-plane around the center except right at the center, where it is either pointing out of the image plane or into the image plane.*

Thin cylindrical magnetic discs of a certain aspect ratio have a special magnetic configuration which is called *vortex state* (compare [Höl04]). The magnetization is following the rim of the cylindrical disc, so that the stray field energy is zeroed out, except at the center where we find a Bloch-line (compare the last section 3.3 for the definition of Bloch-walls). Without the out-of-plane magnetization at the center of the vortex, the exchange energy would be infinite. This can be easily proven using formula 3.8:

$$E_{ex} = A \int_V (\mathrm{grad}\,\vec{m}(\vec{r}))^2 \, \mathrm{d}V = 2\pi h A \cdot \int_0^R \frac{1}{r} \mathrm{d}r \to \infty \qquad (3.16)$$

with the hight of the disc h and its radius R. The infinity follows from the singularity at $r = 0$. Energetically, the small stray field energy is easily compensating the huge exchange energy.

The sense of rotation is called *helicity*, the direction of the magnetization direction at the center is called *chirality*.

The Shifted Vortex

When an external in-plane magnetic field is applied to a vortex structure, the areas with a magnetization parallel to the external field are enlarged, while the opposite areas become smaller (see fig. 3.8). This vortex shift allows us to calculate the saturation

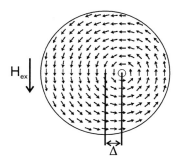

Figure 3.8: *The magnetic vortex structure is shifted to the right by Δ because of the external magnetic field H_{ex}.*

magnetization of a magnetic particle by measuring the shift relative to the external field (see [Höl04] for the derivation):

$$M_S = H_R \left/ \left[\frac{t}{2\pi R} \left(\ln \left(\frac{8R}{t} \right) - \frac{1}{2} \right) - \frac{l_d^2}{R^2} \right] \right. \tag{3.17}$$

with the field H_R at which the vortex core leaves the disc, the radius of the cylinder R, the exchange length l_d and the thickness of the disc t.

If the vortex does not completely vanish but is shifted by Δ by the external field H_Δ, we can assume that

$$\frac{H_\Delta}{H_R} = \frac{\Delta}{R}, \tag{3.18}$$

at least for small shifts $\Delta \ll R$.

As the exchange length is in the range of several nm, l_d^2/R^2 can be set to 0 for cylinders with a diameter of several μm and we finally get

$$M_S = \frac{R H_\Delta}{\Delta} \left/ \left[\frac{t}{2\pi R} \left(\ln \left(\frac{8R}{t} \right) - \frac{1}{2} \right) \right] \right. . \tag{3.19}$$

3.3.2 Magnetic Ripple Contrast

In the inside of magnetic domains, we usually get uniform contrast in TEM Fresnel images (compare chapter 4.2.1) as the magnetic induction is uniform. But some samples show a local fine structure which is called magnetic ripple [Hof65] and produces a stripe contrast in the image. As can be seen in fig. 3.9, the magnetization is following a zig-zag pattern which lowers the magnetostatic energy of the sample as the local magnetizations are partly annihilated. The ripples are causing a locally slightly varying deflection of

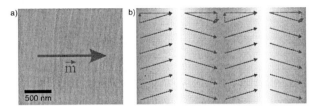

Figure 3.9: *a) Ripple contrast in a TEM Fresnel Lorentz micrograph. b) Magnified scheme of the micromagnetic ripple structure.*

the electron beam inside the TEM. As the component of the local magnetic induction parallel to the mean magnetization is constant, it does not generate contrast in the image. Contrariwise, the perpendicular component with the slight variations does, and thereby the mean magnetization of the domain can be determined as it is perpendicular to the ripple lines.

4 The Transmission Electron Microscope (TEM)

The first transmission electron microscope was built by Max Knoll and Ernst Ruska in 1931. Since then, the TEM has become an important tool for biology, medicine, material sciences and certainly physics. As new operating modes are developed and the resolution limits are steadily lowered by corrective optics, its possibilities are still growing rapidly.

The basic operating principle of a TEM is easy to understand: A specimen thin enough to be transparent for electrons is penetrated by a high energy electron beam, which thereby experiences changes in amplitude and phase by interacting with the specimen. The image is then strongly magnified by a system of magnetic electron lenses (compare [Hur08]) and viewed on a phosphor screen or recorded. The tricky part we have to look at in detail is the contrast formation on the screen or camera, as they can only depict the amplitude of the beam. This leads to difficulties, because a typical TEM specimen is thin enough to be considered as a pure phase object which does not alter the amplitude. Some different techniques which allow the transformation of the invisible phase image into a visible amplitude image as well as the principle design of a TEM are described in the following.

4.1 Basics and Beam Path

Using the example of our FEI Tecnai F30 Regensburg Special, fig. 4.1 depicts the simplified beam path inside a TEM and shows the core components. The *field emission gun* (FEG) emits highly coherent electrons which are focused by a Wehnelt cylinder and accelerated through the anode. The energy of the electrons can be selected in between 50 keV and 300 keV. After the acceleration, a demagnified image of the emitter tip is generated by a condenser system and provides a parallel illumination of the specimen. The beam path below the sample depends on the selected imaging mode, compare the right and left sides of fig. 4.1 for two examples. In *Lorentz mode*, the objective lens is turned off and the Lorentz lens takes its place. As the Lorentz lens if far away from the specimen, the magnetic stray field it produces during operation is close to zero in the specimen plane. The second part of the imaging system and the magnifying projection system produce a real image on the viewscreen or the CCD camera.

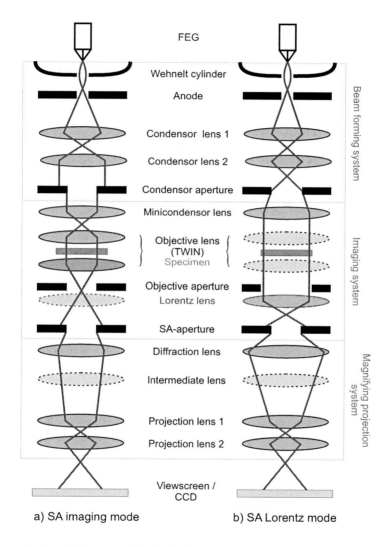

FEG

Wehnelt cylinder

Anode

Condensor lens 1

Condensor lens 2

Condensor aperture

Minicondensor lens

Objective lens
(TWIN)
Specimen

Objective aperture
Lorentz lens

SA-aperture

Diffraction lens

Intermediate lens

Projection lens 1

Projection lens 2

Viewscreen /
CCD

Beam forming system

Imaging system

Magnifying projection
system

a) SA imaging mode

b) SA Lorentz mode

Figure 4.1: *Simplified beam path inside the Tecnai F30 for two special imaging conditions: a) shows the SA imaging mode, b) the SA Lorentz mode. In Lorentz mode, the TWIN-objective lens (the term TWIN emphasizes that a single lens is split into two separated coils) is switched off and the Lorentz lens takes its place. This leads to a nearly vanishing magnetic field at the specimen location, but reduces the maximal spatial resolution of the microscope by about a factor of 10.*

4.2 Magnetic Imaging

Micromagnetic imaging with a transmission electron microscope is a field of research of its own nowadays. From the early techniques like *Foucault* and classical *Fresnel Lorentz* imaging, more advanced ones like *electron holography* and *differential phase contrast* (DPC) have been invented and are being refined more and more (see [Cha84] for details). They make magnetic investigations at lateral resolutions of less than 5 nm possible. It has to be mentioned that both techniques are not limited to magnetic investigations, but their other applications are not the subject of this chapter.

The investigations on GaMnAs and PdFe have been proceeded with Fresnel Lorentz imaging and electron holography, so these two methods will be discussed in more detail.

The basis of all the magnetic imaging modes is that the electron beam, when passing through a magnetic part of the sample, is deviated by the *Lortentz force* $\vec{F}_L = q \cdot \vec{v} \times \vec{B}$ (see fig. 4.2).

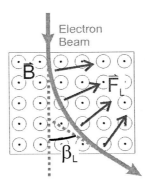

Figure 4.2: *The incident electron beam is deflected by the Lorentz angle β_L in the magnetic field $\vec{B}$ of a specimen. The Lorentz force $\vec{F}_L$ is always perpendicular to the beam.*

4.2.1 Fresnel Lorentz Imaging

The principle of contrast formation in the Fresnel Lorentz imaging mode is depicted in figure 4.3. When the electron beam passes a magnetic area of the sample, it is deflected by the Lorentz angle $\beta_L = \frac{e\lambda}{h} B_\perp d$. e is the elementary charge, λ the electron wave length, d the thickness of the sample, $B_\perp$ the local magnetic induction perpendicular to the electron beam and h is the Planck constant. Parts of the beam which pass differently magnetized areas of the sample are being deflected in different directions which results in diverging or converging partial waves. By increasing the current through the imaging lens, its focal length decreases (the microscopist says it's in overfocus $+\Delta f$) and vice versa. As the image distance s_1 does not change by defocussing, the object distance s_2

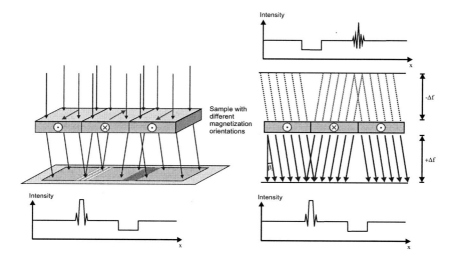

Figure 4.3: *Schematic drawing of the imaging in Fresnel mode (after [Cha84]). By defocusing the sample, magnetic structures can be made visible. By an over- (+Δf) or under-focusing (−Δf), planes below or obove the image plane can be imaged. If the partial waves overlap the intensity increases, if they diverge, the intensity decreases.*

changes according to the thin lens formula $\frac{1}{f} = \frac{1}{s_1} + \frac{1}{s_2}$. This means that a plane below or above the sample is being imaged. As the partial waves converge or diverge in these planes, magnetic domain boundaries can be seen as white or dark lines in the micrograph. By changing from the under- to the overfocus, the contrast inverses which is a quick but not completely foolproof method to distinguish between magnetic and other contrast. As the electron beam in our Tecnai F30 is highly coherent, interference effects at bright domain boundaries are visible quite strongly (see figure 4.4). The fringes appearing at the dark lines because of diffraction at the (magnetic) edge are nearly invisible in daily work. By increasing the defocus more and more, the magnetic resolution achieved by this technique can be increased virtually infinitely, but the decreasing spatial resolution quickly annihilates this advantage.

4.2.2 Electron Holography

As opposed to the classical Lorentz microscopy, the electron holography is an in-focus technique and therefore has a much better lateral resolution. The electron holography utilizes the wave character of the electrons and considering the phase of the beam, the interaction with the sample consists of two mechanisms: The phase of the electron wave

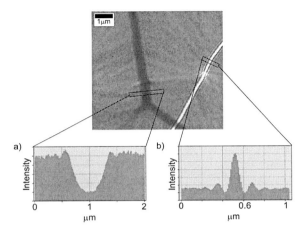

Figure 4.4: *At large defocus values, interference effects can easily be observed in Fresnel micrographs. a) and b) show linescans of a black and a white domain wall.*

changes either because of the B-field of a magnetic sample or because of the reduced wavelength of the electron inside the sample. Like the optical holography invented by Dennis Gabor in 1948, the electron holography records both amplitude and phase of the electron wave. The amplitude is affected by absorption or elastic scattering of the electrons; inelastically scattered electrons all have a different wavelength, so in their sum, they do not produce a clear interference pattern but worsen the signal-to-noise ratio by providing a background illumination.

The experimental setup is as shown in fig. 4.5. The sample has to be placed in the beam in a way that half of the beam can pass undisturbed as reference wave. A *Möllenstedt biprism* (see [Heu05] for details and fabrication), located at the SA-aperture plane between the back focal plane and the image plane, deflects both the object and the reference wave towards each other when charged positively. The overlapping region is magnified by the magnification and projection lenses and recorded by the CCD camera.

In order to get a hologram, the illuminating beam has to be highly coherent. The beam that leaves the beam forming system fulfills this requirement, but in order to get sufficient intensity at the observed area of the sample, the beam is usually focused to a small spot whereby the wave front of the beam becomes spherical. This worsens the coherency significantly. In electron holography, a trick is used that provides both, a sufficient brightness and coherence: The beam is deformed by the condensor stigmator to an elliptical shape (see fig. 4.6). Thereby, the coherence perpendicular to the biprism still exists. Coherence parallel to the biprism is not necessary, as the two beams do not

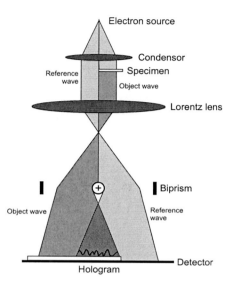

Figure 4.5: *Beam path of the electrons acting as object- and reference-wave. Both waves are deflected by the bipism and interfere in the detector plane.*

interfere in that direction.

The phase shift caused by the magnetism of the specimen can be written as

$$\Delta\varphi_{mag} = -2\pi\frac{e}{h}\Phi = -2\pi\frac{e}{h} \cdot \int \vec{B}\,\mathrm{d}\vec{S}, \qquad (4.1)$$

with the elemental charge e, the Planck constant h, the magnetic field $\vec{B}$ and the area $\vec{S}$.

Combining the magnetic phase shift $\Delta\varphi_{mag}$ with the not further discussed electrostatic phase shift $\Delta\varphi_{el}$ to $\phi = \Delta\varphi_{mag} + \Delta\varphi_{el} + \phi_0$ (ϕ_0 is the incoming wave phase), the recorded intensity can be denoted as follows:

$$I_{hol}(\vec{r}) = 1 + A^2(\vec{r}) + 2\mu A(\vec{r})\cos(2\pi\vec{q_c}\vec{r} + \phi(\vec{r})). \qquad (4.2)$$

$|\vec{q_c}| = \vec{k}\beta$ denotes the carrier frequency of the interference fringes (with β as the overlapping angle) and μ their contrast:

$$\mu = |\mu_{coh}||\mu_{inel}||\mu_{inst}|MTF; \qquad (4.3)$$

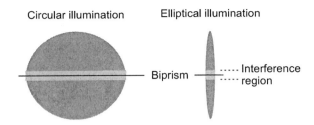

Circular illumination   Elliptical illumination

Biprism   ----- Interference
           ----- region

Figure 4.6: *In order to get both a high intensity and a good coherence perpendicular to the biprism, the usually circular electron beam is deformed elliptically (after [Heu05]).*

it consists of the source coherence (spatial and temporal coherence), inelastic interactions, instabilities and the modulation transfer function MTF of the detector. Both amplitude $A(\vec{r})$ and phase $\phi(\vec{r})$ are encoded as contrast modulation.

In order to get a visible image of the specimen's features, the recorded hologram has to be reconstructed. This is done by a specialized software package which at first calculates the fourier transformation of the micrograph:

$$\text{FT}\{I_{hol}(\vec{r})\} = \underbrace{\text{FT}\{1 + A^2(\vec{r})\}}_{\text{centerband}} + \underbrace{\mu\text{FT}\{A(\vec{r})e^{i\phi(\vec{r})}\} \otimes \delta(\vec{q} - \vec{q_c})}_{\text{sideband +1}} + \underbrace{\mu\text{FT}\{A(\vec{r})e^{-i\phi(\vec{r})}\} \otimes \delta(\vec{q} + \vec{q_c})}_{\text{sideband -1}}$$

$$(4.4)$$

The centerband only contains the diffractogram of a conventional electron micrograph, whereas each of the two sidebands contains the complete complex wave function. One of the sidebands is now cut out and centered in fourier space. By an inverse fourier transformation of this sideband, we get the reconstructed image wave

$$b_{rec}(\vec{r}) = \mu A(\vec{r})e^{i\phi(\vec{r})}. \qquad (4.5)$$

The amplitude $A(\vec{r})$ and phase $\phi(\vec{r})$ are extracted and displayed separately. As the electron beam is far away from having a perfect phase front, the software package has a function which allows the correction of these distortions by taking a so called *reference hologram* in an empty area (where the beam is not altered by the sample) immediately after the normal hologram. This allows the program to strip out the disturbances.

4.3 Generating a Magnetic In-Plane Field by Tilting the Sample

While the main reason for the use of a TEM with a special Lorentz lens is to avoid magnetic fields at the sample location, it is sometimes necessary to generate a field of desired amplitude and direction for micromagnetic experiments. The best way to do this is to used a special sample holder with integrated coils, of which various models are available in our group. But as the experiments in this work are performed at temperatures below room temperature, only the cooling holders can be used, and no cooling holder with integrated magnetic field generation exists so far. Nonetheless, it is possible to use the objective lens and the minicondensor lens which are usually switched off in Lorentz mode for our purpose.

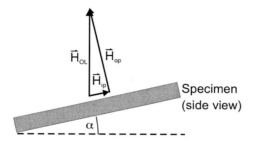

Figure 4.7: *A specimen is tilted by the angle α in the magnetic field of the objective lens $\vec{H}_{OL}$ (or mini condensor lens). This generates an in-plane field $\vec{H}_{ip} = \sin\alpha \cdot \vec{H}_{OL}$ and a remaining out-of-plane field of $\vec{H}_{op} = \cos\alpha \cdot \vec{H}_{OL}$.*

Both lenses produce magnetic fields with a direction perpendicular to the sample plane which makes it necessary to tilt the sample holder in order to get a component parallel to the specimen plane which is what we need e.g. for shifting magnetic domains. Fig. 4.7 depicts the geometry. Tilting the sample by an angle of α results in an in-plane component of the magnetic field of

$$\vec{H}_{ip} = \sin\alpha \cdot \vec{H}_{OL} \,. \tag{4.6}$$

The main disadvantage of this technique is that there is still a strong out-of-plane component of the lens field ($\vec{H}_{op} = \cos\alpha \cdot \vec{H}_{OL}$) which has to be taken into account when interpreting the results of magnetic experiments. In addition to that, the in-plane field can only be applied in one and the opposite direction in respect to the sample if there is only a single tilt sample holder available. If a different direction of the external in-plane field is needed, the sample has to be removed from the sample holder and reoriented. As only the horizontal projection of the specimen can be viewed on the luminescent screen or recorded by the cameras, applying a magnetic field with this method always results in a distorted image of the sample, as the direction perpendicular to the rotation axis is compressed.

Details on the magnetic field generated by the lenses and the precise measurement of the tilt angle α are given in chapters 5.3 and 5.4.

4.4 Diffraction Contrast

While the former sections explained the contrast formation for magnetic samples, the mechanism for imaging the crystallographic structures of a sample is fundamentally different. The technique of diffraction contrast has been used for the experiments with Nb_3Sn in chapter 8. More details on the technique and its applications can be found in [Zwe92] and [Rei97].

As we can see in fig. 4.8, the electrons are diffracted on lattice planes fulfilling the Bragg-condition

$$n\lambda_{el} = 2d \cdot \sin \theta \qquad (4.7)$$

at an angle of 2θ in respect to the incident beam. Hereby, λ_{el} is the electron de Broglie-wavelength, d the lattice spacing and $n \in \mathbb{N}$. As the objective lens – like all magnetic

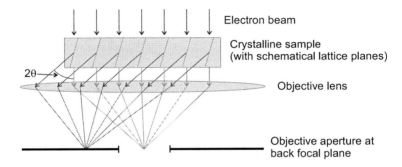

Figure 4.8: *Sketch of the Bragg-diffraction in a crystalline sample. The beams which are diffracted at the same angle are focused to the same spot by the objective lens. By the objective aperture, beams not emerging from the sample at a certain angle can be blocked. After [Zwe92].*

lenses – acts as a collecting lens and thus the not deviated beams are focused on the optical axis on the back focal plane while the diffracted beams are focused on a different point at a distance of $k \sim \frac{2\theta}{\lambda_{el}}$ to the central spot. With the size and position of the objective aperture, we can select which beams shall contribute to the image and which are absorbed by the aperture. Depending on the position of the aperture, we differentiate between the *bright field mode* and the *dark field mode*.

Bright Field Mode

The imaging condition in which the objective aperture is centered around the central beam is called bright field mode. The diameter of the aperture has to be chosen according to the experimental requirements. For now, let's assume that it is big enough so that the central beam can pass undisturbed but that the deviated beams are blocked completely.

As we can see in fig. 4.9 a), those parts of the sample where the beam is diffracted after the Bragg condition are imaged dark as the electrons passing through that area are absorbed by the aperture. Contrariwise, those areas of the sample where the Bragg condition is not fulfilled appear bright in the image.

By increasing the diameter of the aperture, crystallites with larger absolute values of the Bragg-angel appear bright in the image as the beams diffracted by them can pass the aperture.

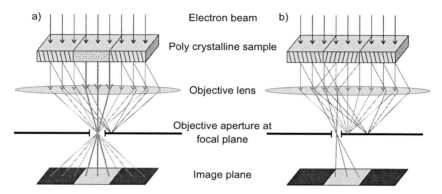

Figure 4.9: *Beam path for a polycrystalline sample in a) bright field mode and b) dark field mode.*
a) The electrons which were diffracted by the crystallites are blocked by the objective aperture and do not contribute to the image. The areas where no diffraction has occured show up bright in the image.
b) All the not diffracted beams and most of the diffracted beams are absorbed by the aperture. Only those beams which are deflected at a certain angle and direction can pass and appear imaged bright. After [Zwe92].

Dark Field Mode

The dark field mode uses the same experimental setup as the bright field mode explained before, with the one difference that the objective aperture is placed beside the optical axis.

This situation is shown in fig. 4.9 b). The non diffracted beams and the majority of the diffracted beams cannot pass the aperture. Only those electrons who are diffracted at a certain Bragg-angle and azimuthal direction are selected. The difference to the bright field mode thus is, that one single crystallite orientation in respect to the electron beam with a range depending on the diameter of the aperture can be highlighted in the image.

4.5 Requirements for a TEM Specimen

The fabrication of samples of high quality is the key to obtain a convincing TEM micrograph. For each material, a different preparation technique has to be used and sometimes none of the established procedures works satisfyingly. Whereas PdFe is on the easy side as it grows seamlessly on commercially available Si_3N_4-membranes, GaMnAs has to be grown on a GaAs substrate and prepared to electron transparency afterwards. The different techniques are discussed in the experimental chapters, but the requirements for a suitable specimen for transmission electron microscopy are the same for each material and are summarized in the following.

- First of all, the sample must have an area with a thickness $< 500\,\text{nm}$, depending on the material and the desired imaging mode in order to be transparent for the electron beam. If micrographs at a high resolution are desired, the thickness has to be even smaller by a factor of 10 or more. The thinned area has to be large enough to cover all the expected features, and thus obviously has to be larger for e.g. magnetic investigations compared to high resolution imaging of crystallographic structures.

- The shape of the specimen has to be roughly circular with a diameter of $2-3\,\text{mm}$. The rim of the specimen has to withstand a significant amount of mechanical stress during the fixation in the sample holder either by spring loaded clamps or screws.

- A (weak) electrical conductivity is a must for TEM samples, as they otherwise accumulate electrical charges from the electron beam and distort the image. For insulating materials, a thin coating of carbon is usually applied.

- Samples investigated at temperatures different to room temperature must have a good thermal contact with the sample holder.

- Magnetic samples have to show a net magnetization perpendicular to their surface as the magnetic imaging techniques are based on the Lorentz force which is zero for a magnetization parallel to the electron beam (compare chapter 4.2.1).

5 Measurement Equipment and Low Temperature Sample Holders

The Tecnai series microscopes are not built for experiments requiring sample cooling, so specialized sample holders which cool the sample and stabilize its temperature have to be used. As the use of these holders requires specialized procedures and as they have limitations in their applications, they are introduced in detail in the first sections of this chapter. After that, new measurement devices for the tilt angle and the magnetic field inside the TEM are presented.

5.1 Philips Double Tilt Liquid Nitrogen Cooling Holder

The Philips double tilt liquid nitrogen cooling holder in principle provides a temperature range of 93 K up to about 100 K above room temperature, but there is a gap where the temperature cannot be stabilized. The holder is connected to a manual control unit which measures the temperature with a sensor placed a few centimeters away from the mounted specimen with a resolution of 1 K. The relation between the indicated temperature and the real temperature of the sample has to be verified for each specimen as the specimen is heated up by the electron beam by an unknown amount because the thermal conduction and other physical properties which affect the heating vary from sample to sample.

When liquid nitrogen is filled into the holder's Dewar, the temperature begins to fall and reaches the lower limit after roughly half an hour. If a higher temperature is necessary, the desired temperature can be set by counter-heating the tip of the sample holder with a constant current source included in the control unit which powers a heater filament inside the holder. Each heating current leads to an equilibrium temperature, but as the temperature is reached asymptotically, it is never really stable during the experiment. Setting the appropriate current requires a lot of practice. As the maximum heating current the manufacturer allows is 700 mA, the temperature range between about 140 K and room temperature cannot be set.

Experiments with this holder have no time limit, as the liquid Nitrogen can be topped up easily. The Nitrogen in the holder's Dewar is constantly bubbling which makes the specimen move around and makes the use of electron holography impossible.

5.2 Gatan Single Tilt Liquid Helium Cooling Holder

For temperatures below 93 K, a liquid Helium cooled sample holder has to be used. This special holder was manufactured by Gatan and can reach a minimum indicated

Figure 5.1: *The liquid helium holder inserted in the TEM. (1) compu-stage, (2) tilt-sensor, (3) suspension spring, (4) fill in nozzle, (5) He exhaust, (6) Dewar.*

temperature of about 12 K.

Temperature control can be done in two different ways: First, the gas flow of the evaporated helium out of the Dewar – and thus the pressure dependent temperature of the liquid helium – can be regulated. This makes it possible to manually set the temperature at the specimen location between 12 K and about 50 K. The second method is to use the microcontroller equipped control unit which operates a heater filament inside the sample holder. The desired temperature value is entered into the controller which adapts the heater current. As the equilibrium point is very delicate, a temperature change leads for several minutes to a swinging of the temperature around the desired one with amplitudes up to 5 K. After several minutes, the temperature is stable to about 0.5 K.

The maximum time for which the 12 K can be reached is about one hour, e.g. 25 K set by a lower gas flow is stable for one and a half hour. As counter-heating leads to an increased evaporation rate of the liquid helium, it shortens the time available for performing experiments at low temperatures. Toppig up the Dewar with liquid Helium does not work most times because the filling process disturbs the internal gas flow of the sample holder. As one complete fill needs about 30 l of liquid Helium (which costs > 200 EUR) which cannot be recollected, topping up should not be tried and thus the time for the experiment is strictly limited. The filling up of the holder takes around 1.5 hours and the heating up to room temperature after the experiment takes 2-3 hours.

As with the liquid nitrogen holder, the problem with the bubbling cooling liquid is present with the liquid helium holder, too, so no proper electron holography is possible with this holder.

The second major problem with this holder is that its rear end with the Dewar is too heavy for the Tecnai's compustage. Gatan solves this problem by suspending the holder with two springs whose strength can be adjusted by two screws (see fig. 5.1 (3)). As a result of its weight and the springs, the holder does not exactly follow the α-tilt[1] of the stage. Even when α is set to $0°$, the holder is noticeably tilted. As the knowledge of the precise angle between the electron beam and the sample plane is essential for calculating the in-plane magnetic field the sample sees (compare chapter 4.3), the angle of the holder has to be measured externally. The following paragraph introduces a measurement device especially designed for this application.

5.3 Helium Holder α-tilt Measurement

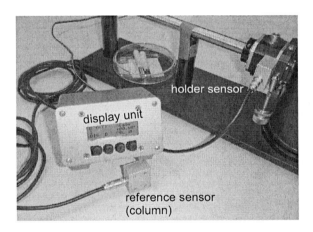

Figure 5.2: *The liquid helium holder equipped with the tilt sensors. The reference sensor has to be mounted to the microscope column.*

The developed device uses two accelerometers which measure the gravitational force along one of their axis. Thereby, the relative tilt-angle can be calculated: $\alpha = \cos^{-1} G$ (G is the measured acceleration). Two sensors are needed because the inclination of the microscope column itself is varying with time as the column is air suspended. The reference sensor is mounted on the microscope column on the opposite side of the stage, where a plate coplanar to the plane of the α-tilt can be found; the second sensor is mounted to the sample holder (see fig. 5.2). A microcontroller unit calculates the relative angle with a precision of $\pm 0.1°$ and displays both the angle and, if the actual

---
[1]The α-tilt is the tilt of the sample holder around its axis.

lens excitation is entered, the calculated in-plane magnetic field corresponding to the relative α between microscope column and specimen plane.

5.4 Precise Magnetic Field Measurement at the Sample Location

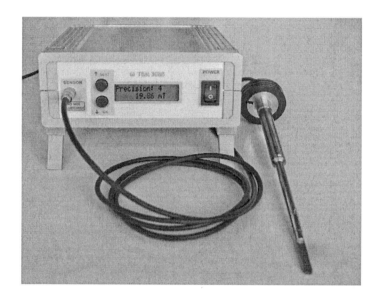

Figure 5.3: *The magnetic field measurement device with the microcontroller equipped supply- and acquiring unit on the left and the optimized sample holder with the built-in hall sensor on the right.*

Knowledge of the precise value of the magnetic field generated by the objective- and mini-condensor-lens is essential for most TEM measurements with magnetic samples. [Ott01] describes a simple set-up using a sample holder with a built-in hall probe, a constant current source and a voltmeter. This experiment can deliver an accuracy of about $\pm 2\,$mT which is sufficient for most applications. But for soft magnetic materials like (Ga,Mn)As ad PdFe, especially around their Curie temperature, the magnetic field has to be determined more precisely.

The set-up was improved as follows: As a first step the tip of the sample holder where the hall sensor is mounted is now made of titanium as it has been discovered that the formerly used brass had a small magnetic remanence because of impurities. Second, a special circuitry with a highly stabilized constant current source and a high sensitivity ADC[2] has been developed which allows the calibration of the sensor's specific amplification factor, compensates the earth magnetic field and is able to automatically average over predefined numbers of data points (see fig. 5.3). The accuracy could thereby be improved by at least a factor of 10 leading to a resolution of better than ± 0.2 mT.

Fig. 5.4 shows the measurement of the magnetic fields in our Tecnai transmission electron microscope at 300 kV acceleration voltage. The linearity is quite good up to $\pm 50\%$ lens excitation, where the saturation effects of the pole pieces begin to show up. The remaining field of 0.4 mT at 0% excitation of the objective lens can be eliminated by exciting the mini condensor lens by -30%; the accompanied image distortions are small and can easily be corrected. Lower acceleration voltages reduce the magnetic field needed for the electro-magnetic lenses because of the lower relativistic mass of the electrons, so the magnetic field characteristic has to be measured for each desired acceleration voltage.

---

[2] Analog-to-Digital-Converter

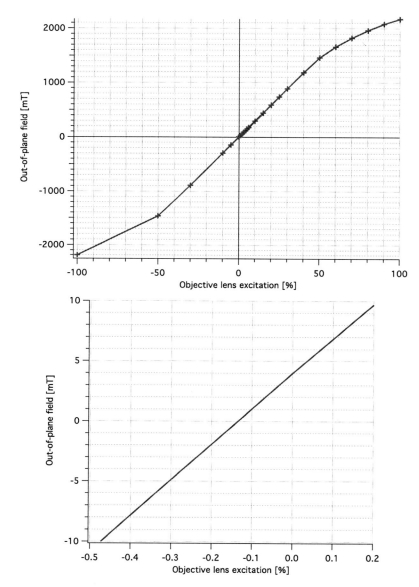

Figure 5.4: *Magnetic field perpendicular to the sample plane generated by the objective lens at 300 kV (the mini-condensor lens was not excited). The top graph gives an overview over the whole range while the lower one is an enlarged view of the area around 0 mT. The field at 0% excitation is 0.4 mT. For the negative excitations, the objective lens had to be commutated, see [Hur08] for details.*

6 GaMnAs

Since H. Ohno first reported magnetism in Mn-doped GaAs in 1996 [Ohn96], it has become the most prominent and best known representative of the magnetic semiconductors. The TEM investigations on (Ga,Mn)As this chapter is about form the project B2 of the Collaborative Research Area 689 entitled "Spin Phenomena in Reduced Dimensions". The aim of this project is to investigate the micromagnetic domain structures and the magnetic phase transition of (Ga,Mn)As samples grown at the University of Regensburg.

The first section introduces (Ga,Mn)As and explains the origin of its magnetism as well as how the magnetic properties can be altered. This part draws largely on [Lan06], where more details can be found. In the following, the different preparation methods evaluated for (Ga,Mn)As are presented and their applicability is rated. The performed microscopic experiments are summarized in the last part.

6.1 Physical Properties

6.1.1 Magnetic Properties

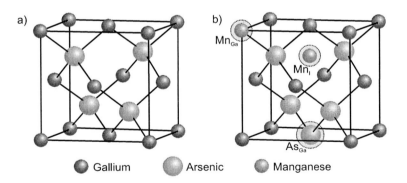

Figure 6.1: *a) Zincblende crystal structure of pure GaAs. b) Crystal lattice of (Ga,Mn)As with Mn on Ga sites (Mn_{Ga}), Mn interstitials (Mn_I) and As antisites (As_{Ga}).*

(Ga,Mn)As is based on the III-V semiconductor GaAs with zincblende structure (see fig. 6.1). By substituting some Ga atoms with Mn atoms, we get $Ga_{1-x}Mn_xAs$; x denotes the part of the Ga atoms which are substituted. At first, we focus on just this case, ignoring the existence of all other lattice defects. $Ga_{1-x}Mn_xAs$ becomes ferromagnetic when $x \approx 0.012$ or more (see [Ohn99]); a mostly homogenious sample can be grown by molecular beam epitaxy (MBE) up to $x \approx 0.1$. As the highest Curie temperatures are reached at about 5-6% Fe [Sad06a], all our samples have a concentration in that range. In order to understand the formation of the magnetism in (Ga,Mn)As, we have to regard the electronic structure of Mn first.

The K- and L-shells are completely occupied with 10 electrons, the M-shell has 2 electrons in its 3s-orbital, 6 in the 3p-orbital and 5 in the 3d-orbital. The 4s-orbital of the N-shell is completely filled, so the electron configuration is

$$1s^2 2s^2 2p^6 3s^2 3p^6 3d^5 4s^2.$$

As the 3d orbital momentum $L = 0$, the five 3d electrons lead to an overall spin of 5/2. When acting as a dopand in GaAs, Mn can have three theoretically possible electronic configurations when situated at a gallium site:

- The most prominent model is a Mn^{2+}-ion in $3d^5$ configuration with a spin of $S = 5/2$. As both 4s-electrons of the Mn are necessary for the binding, the Mn is double positively charged. At the lattice position of the trivalent gallium, it acts as an acceptor, generating one hole per Mn-ion and making the sample p-doped. This state is usually labeled $3d^5 + h$, where 'h' stands for 'hole'. If all the Mn atoms would be in that state, the p-doping would be equal to the Mn concentration x. But as p has always been measured lower than x, there must be other additional manganese configurations.

- Manganese could also act as an ionized acceptor if one of the electrons is bound to the d-shell and thereby the three bindings of the replaced Ga atom are being saturated by this atom and the two 4s-electrons. Such a Mn atom would not contribute to the p-doping.

- The third possible configuration is the neutral state of Mn (A^0) with $3d^4$-configuration, labeled as $A^0(d^4)$. This state could be verified together with the $A^0(d^5+h)$ state by [Twa99]. The three bindings of the replaced Ga are saturated by the 4s-electrons and one d-electron, which leads to a vacancy in the d-shell and the remaining 4 electrons result in an overall spin of $S = 2$ [Sap02]. This configuration is very seldomly reported and presumably strongly dependent on the growth parameters.

The most dominant electron configuration is the first mentioned $3d^5 + h$ configuration, which has been verifyed by many groups. But as not all of the Manganese atoms are in that configuration and the others do not contribute to the p-doping, it is always slightly smaller than the Mn concentration.

As already depicted in fig. 6.1, there are also crystallographic defects related to the Mn-implantation:

- It is possible that a manganese atom is situated neither on a Ga nor a As site, which is then called manganese interstitial (Mn_I). This results in the behaviour of the Mn as a double donor which reduces the p-concentration by compensating the holes with electrons. As the magnetic interaction in (Ga,Mn)As is carrier mediated, this decrease in the free carrier concentration has negative effect on e.g. the Curie temperature. But this is not the only effect reducing T_C as the Mn_I are coupling antiferromagnetically to the next Mn_{Ga} acceptors and therefore both Manganese atoms do no longer contribute to the ferromagnetic order of the sample.

- The growth of (Ga,Mn)As has do be performed with an excess of Arsenic which results in As atoms situated at Ga sites, called *arsenic antisites* As_{Ga}. Those antisites also lead to a compensation of holes [Woj04], which reduces T_C.

6.1.2 Influence of Thermal Stress

The distribution of the beforehand mentioned configurations of (Ga,Mn)As can easily be altered by moderate heating. When annealing the sample at temperatures close to the growth temperature ($180 - 250\,^\circ$C), Mn_I atoms are moving out of the (Ga,Mn)As and the Curie temperature is significantly improved (see [Yu02], [Edm04], [Wan04], [Ade05], [Sta05]).

Unfortunately, T_C does not stay at this high level when all the Manganese interstitials have moved out and heating is continued. Instead, T_C is going down again noticeably ([Ade05], [Sta05], [Pot01]). If the temperature is higher than the growth temperature, the decrease in T_C is even steeper and can be reduced e.g. by 20 K in three hours [Sta05]. The reason for that is that the Mn_{Ga} are forming clusters, which decrease T_C ([Xu05], [Rae05]). The Mn clusters mostly consist of Mn atoms surrounding one As atom. The larger the cluster, the more localized is the hole distribution around the center As atom because the fermi level splits increasingly, but this located hole concentration does not contribute to the ferromagnetic coupling and thus reduces the Curie temperature.

6.2 Sample Preparation

The preparation of proper TEM samples out of (Ga,Mn)As grown on a GaAs substrate cannot be done in the usual way because of some of its pecularities. First to mention is the high sensitivity on the temperature as shown above. The standard TEM preparation method for bulk samples is a combination of mechanical thinning and ion etching, whereas the last step heats up the sample significantly as the next section shows. The second attribute of (Ga,Mn)As is its extreme brittleness which is even pronounced by the fact that the area of intrest of TEM specimens has to be electron transparent with a thickness of less than 500 nm and hence is even more prone of breaking loose. The different strategies for getting working specimens are described in the following.

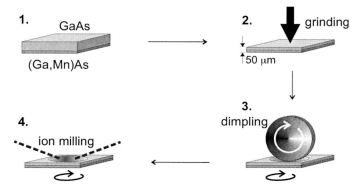

Figure 6.2: *1. A piece of 2x2 mm is cut of out of the waver.*
2. The piece is grinded to a thickness of approx. $50\,\mu m$, removing parts of the GaAs substrate.
3. A grinding wheel makes a dimple in the center of the rotating specimen. The thinnest spot in the center has now a thickness of $15 - 20\,\mu m$.
4. Material is slowly removed by two ion beams until the center of the sample is perforated; finally the sample is polished with very low energy ion beams.

6.2.1 Dimpling and Ion Etching

The most common specimen preparation technique for transmission electron microscopy is the mechanical thinning of a small piece of the sample with subsequent ion etching [Kai00]. A 2x2 mm^2 small piece is cut out of a GaAs waver with the (Ga,Mn)As layer grown on thereupon by MBE. The piece is then mechanically grinded with a diamond disc down to a thickness of about $50\,\mu$m. The resulting foil is further thinned by carefully polishing a dimple on its GaAs side using a Gatan dimple grinder. This step has the advantage, that only a small area in the center is really thin (about $15-20\,\mu$m), while the thicker rest of the sample stays relatively thick and reassures the mechanical stability. The final thinning is performed afterwards by a two-step ion etching, using either a Gatan Precision Ion Polishing System (PIPS) with argon gas as etchant or a Balzers Baltec with xenon gas. In the first step, the sample is etched using an acceleration voltage for the ions of $\leq 3\,$kV for several hours until a small hole appears in the center of the dimple. In the second step, the surroundings of the hole are polished with a reduced acceleration voltage of $1.5\,$kV or less in order to receive an even surface and reduce defects introduced by the high energy ion etching. All these steps are illustrated in fig. 6.2. In the following, only the Gatan PIPS is used.

The main disadvantage of ion etching is that the sample is heated up by the ion bombardement. The degree of heating during ion milling in a PIPS was not known and has therefore been investigated within the scope of a diploma thesis. A short overview is given hereafter, [Wen08] shows details.

The temperature has been measured using three different approaches: By a thermo-couple glued to a test sample, by a calibrated infrared thermometer and by special thermo-lacquers.

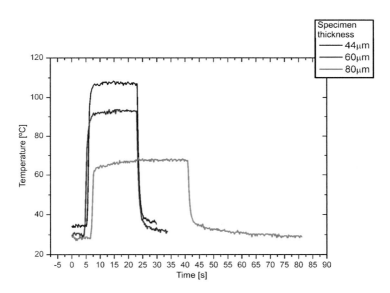

Figure 6.3: *Impact of the sample thickness on the temperature, measured with an infrared ther-mometer. Three GaAs samples of different thicknesses have been ion milled with the same parameter set (acceleration voltage 4 kV, etching angle 10 °). The temperature rise is bigger for thinner samples. Taken from [Wen08].*

- A **thermo couple** made of 70 μm thick wires is glued on the back side of the sample. This method is the less accurate one, first, because it averages over a big part of the sample, and second, because a significant mass and an additional thermal leakage is added to the sample.

- An **infrared thermometer** is mounted above a special IR-transparent sight glass. The focus of the thermometer optics can be adjusted to a diameter of 0.8 mm which is already significantly better compared to the thermo couple, but the focus point cannot be adjusted precisely enough to cover only the electron transparent thinnest part of the sample.

- The sample is covered with a thin layer of a special **thermo lacquer**. Those laquers change their macroscopic structure irreversibly at a predefined temperature and are commercially available. As the minimum achievable thickness of the lacquer is $20 - 30\,\mu\text{m}$, it adds significant mass to the sample itself. The two lacquers used had a transition temperature of $79\,^\circ\text{C}$ and $169\,^\circ\text{C}$, respectively, so only this two temperature limits could be determined by this method.

The results of all the measurements - no matter which technique had been used - can only give a lower limit to the real temperatures reached during ion milling as even the thinnest sample investigated in [Wen08] is by at least a factor of 100 thicker than the sample areas observed in the transmission electron microscope. Fig. 6.3 shows the dependence of the sample temperature during ion milling on the thickness. Different settings of the ion mill resulted in measured temperatures of up to $300\,^\circ\text{C}$.

The PIPS ion mill was recently equipped with a liquid nitrogen cooling unit which lowers the sample temperature around 115 K and low voltage ion guns which make it possible to use acceleration voltages down to 0.1 kV. The combination of this two improvements, of which the latter one prolongs the thinning time significantly, may allow to use ion milling even for temperature sensitive samples.

6.2.2 Splinter Technique

In order to avoid any heating of the sample and nonetheless use standard (Ga,Mn)As samples grown on GaAs substrates, the splinter technique is the only way to go.

For cross-section samples, this technique is performed as follows: A several square millimeters sized piece of the sample is mechanically grinded to a final thickness of $50 - 100\,\mu\text{m}$ and placed between several layers of soft paper, e.g. filter paper. Punctual pressure is applied a few times, until the splinters have an average size of roughly 1 mm. The pieces are then viewed under the optical microscope and those splinters with the sharpest edges are selected. Finally, each one of the good ones is glued to a copper ring with a 90° support using conductive glue in a way that the thinnest part is centered over the hole of the copper ring (see fig. 6.4). This part of the sample should be thin enough to be electron transparent.

The splinter technique has two main advantages: It is a very quick method and it leaves the sample's physical properties unaltered, but its drawbacks cannot be neglected. They are the need of large amounts of raw material as the chance of getting a good sample is low, as well as the small electron transparent region of the sample depending on the splinter angle.

For the sake of completeness, it has to be mentioned that this method can be adapted for plane-view samples, too, but as the success rate in this case is even smaller as are the electron transparent regions, the practical use is infinitesimal.

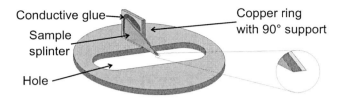

Conductive glue

Sample splinter

Hole

Copper ring with 90° support

Figure 6.4: *Copper ring with 90 ° support. A small splinter is glued to the support with conductive glue. The inset on the right shows a magnified view of the tip of the splinter with the substrate on the left and the evaporated layer on the right.*

6.2.3 Chemical Etching

The last method we used for preparing thin TEM samples of (Ga,Mn)As is based on chemical etching and is also used by [Sug08]. The main disadvantage of this technique is the need of an etch stop layer made of AlAs which is not the perfect surface to grow (Ga,Mn)As on. This is mostly compensated by the growth of a thin layer of GaAs on top of the stopping layer.

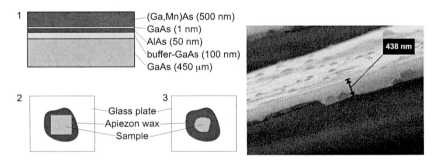

Figure 6.5: *Left: 1: (Ga,Mn)As multilayer for chemical etching with AlAs as stopping layer. 2: Sample glued on a glass plate with Apiezon wax, GaAs facing up. 3: The sides and the rim of the sample are covered with wax.*
Right: REM (Raster Electron Microscope) micrograph of an wet chemical etched GaMnAs layer. The desired thickness was 500 nm.

The complete growth structure for a sample meant for chemical etching is the following multilayer which is deposited on top of a GaAs waver: (Ga,Mn)As (500 nm) / GaAs (1 nm) / AlAs (50 nm) / buffer-GaAs (100 nm) (compare fig. 6.5 (1)). After the growth, the sample is mechanically grinded from the substrate side down to a thickness of 200 μm. Now, the sample is glued with Apiezon wax to a glass plate with the substrate facing upwards and the sides as well as the rim are covered with wax, as shown in fig. 6.5 (2). The covering of parts of the surface is necessary in order to get a mechanically stable rim after the etching which makes it possible to handle the piece.

The substrate and buffer-GaAs is then selectively etched with acetic acid : hydrogen peroxide : water 5:1:5, using the AlAs as a stop layer. Etching at room temperature results in an etching rate slow enough to make it possible to determine the end point visually. So after the first hour, the etching is observed under an optical microscopy and terminated when the AlAs layer is reached. This point can be determined because the reflective properties of the sample change significantly when reaching the stopping layer, because its slower etching rate (about 1/20 of the rate in GaAs) leads to a smoother surface. The time window for stopping the etching is less than a minute which is tiny compared to the overall etching time of 1-2 hours. The right image of fig. 6.5 shows a finished sample.

The remaining glue has now to be removed by acetone. This makes the sample splinter into many pieces, of which the most promising ones are clamped between two TEM sample grids made of copper.

6.3 Thinned Samples

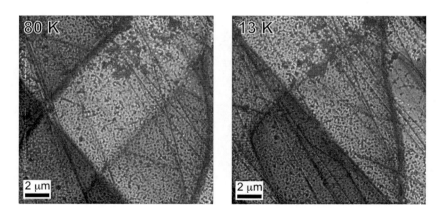

Figure 6.6: *Fresnel Lorentz micrographs of GaMnAs (under-focus, Lorentz-lens excitation 86%), both images show the same area of the specimen. The left image was taken above T_C at 80 K, the rigt image at 13 K well below T_C. The images show bending contours which change with temperature, but no magnetic features are visible as no new image features appear.*

Samples prepared with all the above mentioned methods were investigated in the TEM by Fresnel Lorentz imaging and electron holography. In order to distinguish if visible image features were of magnetic origin or not, micrographs taken at different temperatures but with the identical microscope settings have been compared.

The most promising — because most sensitive — technique [Hur04] for imaging magnetic domains in (Ga,Mn)As was electron holography (compare chapter 4.2.2). Unfortunately, we had to learn that electron holography is completely inexecutable at low sample temperatures as the small bubbles emerging in the liquid helium and possibly the flow of the helium through the capillaries cause a permanent slight movement of the sample. Together with the necessary exposure times for an electron hologram of 1-4 seconds, the interference fringes were disturbed too much and no reconstruction was possible.
So all the experiments had to be performed using the Fresnel Lorentz imaging mode.

None of the samples has shown any magnetic behavior. Fig. 6.6 shows the most promising specimen above and well below T_C which was in the range of 45 K. The small dots are residuals from the etching process, the lines and bands are typical bending contours found in every TEM micrograph of a crystalline sample.

6.4 Stray Field Imaging

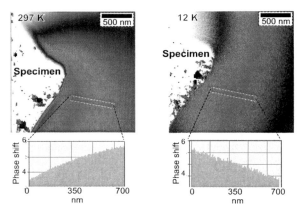

Figure 6.7: *Electron holography stray field micrographs of a piece of (Ga,Mn)As at room temperature and 12 K. The phase shift is color-coded. The line scans below the images show the phase shift in that area. The difference in the stray field is that pronounced that the gradient has even changed its sign between the two micrographs. The contrast inside the specimen which is not electron transparent is an artefact of the reconstruction process.*

As the direct imaging of magnetic domains did not work at all, the idea of detecting the magnetic stray field was developed. This made it possible to use electron holography even at low temperatures, as the weak movement of the sample didn't affect the interference fringes in the vacuum.

The samples for the stray field imaging are simple to produce, as no thinning is required because electron transparency is not necessary. Hence, the only thing to do is to glue a small piece of the processed wafer with a thick layer (for a strong magnetic signal) of (Ga,Mn)As on a copper ring with conductive glue (compare the top of fig. 6.9).

The first experiments were very promising, strong variations in the electron beam phase were detected around the sample which changed their shape with temperature, see fig. 6.7. Surprisingly, the stray field changed significantly even for temperatures well above T_C and was still present even at room temperature. A control experiment with completely unmagnetic copper replacing the (Ga,Mn)As piece showed similar stray fields. The most probable reason for this effect is the following: In order to get the for

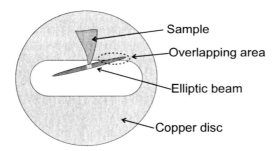

Figure 6.8: *Schematic view of the elliptically deformed beam used for electron holography compared to a copper specimen ring. Depending on position of the sample, parts of the beam hit the ring and / or parts of the sample.*

electron holography necessary spatial coherence of the electron beam perpendicular to the biprism, the beam has to be reshaped elliptically by using the condensor stigmator lenses. As shown schematically in fig. 6.8, this usually leads to a situation in which parts of the sample, which can be far away from the observed region, are hit by the electron beam. Every time when the region of interest changes and the sample is moved, which is particularly the case when recording an empty reference hologram, the illuminated area at the tails of the beam changes. Because of the finite resistance of the copper ring and inevitable surface residues, the specimen slightly charges up, depending on the location and the size of the area illuminated by the electron beam.

This usually does not play a role in electron holography, as the object hologram and the reference hologram are taken close together at distances of merely several μm, but in our case, where the stray field shall be imaged, it is necessary to take the reference hologram as far away from the sample piece as possible in order to be out of the reach of the stray field.

A second effect playing an important role when comparing holograms taken at different times is that the imaging conditions of the microscope are not completely stable over time. This can be shown by a simple experiment: You take three holograms of a completely empty region of the sample, so to say of the vacuum, the first two in quick succession, but the third one after a delay of several minutes. When reconstructing the second hologram using the first one as reference, the final image shows no phase shift. But doing the same with the third hologram results in a visible phase shift pattern, showing phase shifts of π or even more. This tells us that the wave front arriving at the CCD detector has changed significantly over time. Therefore, the object hologram and the reference hologram usually have to be taken within a few seconds. The third possible disturbance arises from the pole-touch detection feature of the compustage of our Tecnai microscope. In order to prevent that the sample holder tip rams the objective lens pole pieces and thereby causes considerable damage to the holder or the microscope itself, a special circuitry is installed which immediately stops the movement of the sample holder when it gets into contact with one of the pole pieces. This is done by detecting an electrical current between the lens and the holder which makes it necessary to keep them on a different potential which is 2 V for the Tecnai microscopes. The irregular shape of the sample holder tip and its varying relative position to the pole pieces produces a weak changing inhomogeneous electrical field. While this field is usually far to low to be noticed, it has to be cancelled out in order to achieve the maximum possible sensitivity for our stray field measurement. This was done by adding a switch to the microscope with which the sample holder can be disconnected from the 2 V supply and connect it to the microscope ground instead. The detection must only be switched off when the stage is not moving at all in order to prevent heavy damage.

In order to circumvent the electrical charging effect while at the same time paying attention on the temporal instabilities, the special image recording and reconstructing procedure shown in 6.9 was developed. For one final micrograph showing the difference in the stray field between two different temperatures, four holograms have to be taken - one object (sample) hologram and one reference hologram at each temperature. The important thing is the reconstruction process, which allows us to correct both disturbing effects: The reference hologram taken at the low temperature is reconstructed using the reference hologram taken at room temperature (as the highest temperature achievable with the liquid helium cooling holder) as reference. This results in a phase pattern corresponding to the changes in the imaging conditions caused by the instability of the microscope over time. In a similar way, the object hologram taken at the low temperature is reconstructed using the object hologram at room temperature as reference. This gives us the changes in the expected stray field itself plus the disturbances coming from the optical system. When one resulting image is subtracted from the other, only the pure stray field remains.

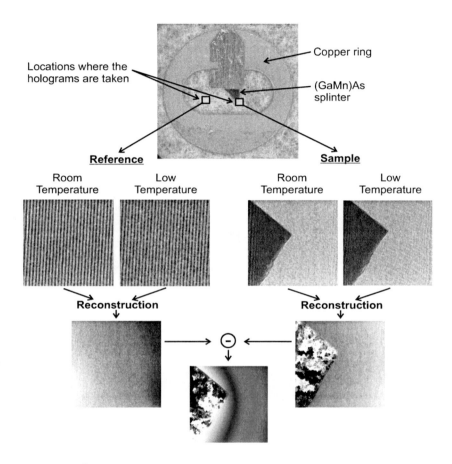

Figure 6.9: *Scheme of the recording and reconstruction process for disturbance corrected holograms at different temperatures.*

The topmost image shows the locations of the reference and object holograms. For each temperature, one hologram is recorded at both places. The reconstruction is done in two steps: The first is the separate reconstruction of the reference and object holograms using one of the pair as reference. The second step is to subtract the obtained images from each other.

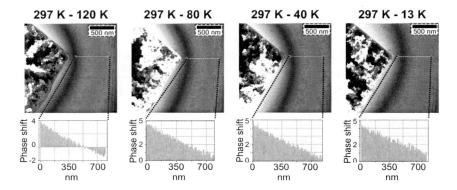

Figure 6.10: *Stray field images generated by the procedure described in section 6.4 with a reference temperature of 297 K. The patterned triangle on the left of the images is the specimen, the pattern is an artefact and has no physical meaning. The color-coded phase shift pattern is roughly the same for all four temperatures although the transitions temperature lays with 63 K between the second and third micrograph. The linescans below the images show the gradients in the same areas of the images, its slopes do not change significantly.*

Fig. 6.10 shows some representative results of this procedure. On the left side of the micrographs, a corner of the (Ga,Mn)As sample is visible. The pattern inside is random and an artefact of the reconstruction process as no intensity is transmitted in this area. This sample had a Curie temperature of T_C=63 K, so the first two micrographs should show a different phase pattern outside the sample compared to the third and fourth one. As the color-coded phase shift and the linescans depicted below the micrographs show roughly the same gradient, no significant change could be measured. The existing phase pattern is temperature independent and therefore not of magnetic but electrostatic origin.

6.5 Conclusion

Despite the fact that no magnetism could be detected for (Ga,Mn)As samples with our Tecnai microscope, the research in this field led to important results.

- The difficulties experienced while preparing samples of (Ga,Mn)As and their solutions improved our knowledge of sample preparation techniques already in use like e.g. ion milling and new techniques were introduced, like chemical etching and the splinter technique.

- The diploma thesis of Martin Wengbauer [Wen08] was initiated by the research on (Ga,Mn)As and provides useful indications on the thermal stress during ion milling.

- The single components that possibly disturb the electron holography have been investigated in detail and new techniques have been developed which solve these problems.

The reasons why no magnetic behavior of the investigated (Ga,Mn)As samples could be observed can not be determined completely. It is possible that the magnetic fields have been to weak to be detected with the used techniques, particularly as electron holography could not be performed on electron transparent samples because of mechanical instabilities. On the other hand, the real temperature of the part of the sample that is imaged and thereby irradiated by the high energy electron beam is still unknown but subject to current research projects, more details on this in chapter 9. It is also possible that the thermal conductance the used experimental set-up was to bad and temperatures considerably below T_C could not be reached in the case of (Ga,Mn)As.

7 PdFe

$Pd_{1-x}Fe_x$ with an iron concentration of around $x \approx 0.05$ is a favored material for contacting carbon nano tubes (CNT) for spintronic devices. In order to understand the behavior of such spin devices, which strongly depends on the spin injection from the PdFe into the CNT, the micromagnetic properties have to be known. As the Curie temperature of $Pd_{0.95}Fe_{0.05}$ is approximately 170 K and thus well in the range of the LN_2 sample holder, it is a suitable subject for intense magnetic TEM studies.

This chapter first outlines the mechanism of magnetism in PdFe. After the section about sample preparation, it is shown how special specimens can be used to determine the temperature distribution on Si_3N_4-membranes. After this introductory parts, the magnetic experiments performed with PdFe and the results are presented.

7.1 Magnetism in PdFe

Palladium is paramagnetic, but its value of the *Stoner criterion*[1] is $J \cdot n_0 = 0.78$ (n_0 is the density of states at E_F, ferromagnetism for $J \cdot n_0 > 1$), and hence palladium iron is quite close to the ferromagnetism. When being doped with Fe, a concentration as small as $x = 0.001$ of iron is enough for the resulting PdFe to become ferromagnetic at low temperatures (compare [Smi70]).

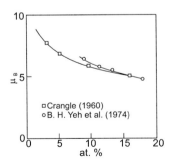

Figure 7.1: *Average magnetic moment per iron atom according to [Yeh75].*

---

[1]Edmund Clifton Stoner (1899-1968) introduced a band model of ferromagnetism

As the mean distance of two Fe atoms is quite large at such low concentrations, a direct interaction can not take place. The commonly accepted theory is that the local magnetic moments of the iron atoms interact with the palladium host matrix, polarizing it and thus induce a magnetic moment in the Pd. This explanation is supported by the fact that the average magnetic moment per Fe atom can reach values measured up to 12 μ_B (μ_B is the Bohr magneton) for $x = 0.025$ which is more than the iron itself can deliver with its 10 d-electrons and 1 μ_B per electron. The dependence of the average magnetic moment per iron atom is shown in fig. 7.1.

The Curie temperature of PdFe strongly depends on the Fe concentration, [Smi70] measured an increase of T_C of 37 K per at. % Fe for concentrations below 5 %, the increase rate is lower for higher concentrations.

7.2 Sample Preparation

The PdFe investigated in this work has been grown directly on commercially available silicon nitride (Si_3N_4) membranes. They consist of a quadratic silicon substrate with an edge length of roughly 2.1 mm. In the center, there is a 100 μm x 100 μm window which is covered by 35 nm of Si_3N_4. When a patterned sample is desired, structuring has to be performed before the PdFe is evaporated. This is done in the following way:

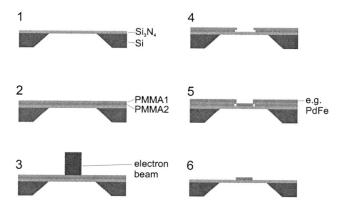

Figure 7.2: *Preparing a Si_3N_4 membrane sample. 1: Cross-section through the center of a Si_3N_4 membrane. 2: Two layers of PMMA are deposited. 3: The microstructure is written with an electron beam which exposes the lacquer and breaks its bonds. 4: The exposed PMMA is removed. 5: Depositing the magnetic material. 6: Removal of the excessive laquer and magnetic material.*

In the first step, the membrane is covered by two layers of *PMMA* [2] which differ in concentration. After that, the desired microstructures are written into the laquer in a REM by electron beam lithography. This breaks the bonds in the exposed areas which can now be removed with MIBK[3] and isopropanol. The two different concentrations of PMMA lead to a so called under cut structure, as the lower layer is dissolved in a slightly larger area than the upper layer. Now, the magnetic material is thermically evaporated onto the sample in an ultra high vaccuum chamber. The under cut profile of the lacquer prevents the evaporated material from getting into direct contact with the lacquer which improves the shape and simplifies the following lift-off. The evaporation has been done by Ondrej Vavra in the group of Prof. Dr. Weiss using their MBE apparatus. The last step is the removal of the excessive material by dissolving the non exposed PMMA with acetone; this process is called lift-off. The single steps are depicted schematically in fig. 7.2.

All our samples have been grown with a desired concentration of iron of 6 at.%. Wu [Wu73] has shown that the Curie temperature is directly proportional to the iron concentration at a rate of approx. 20 K / at.% Fe, so a slight variation of the iron concentration has a large impact on T_C (compare fig.7.3). As there were problems with the palladium oven during the MBE growth of several samples, the real iron concentration varies between 5 % and 8 % which leads to a T_C between 160 K and 220 K.

7.3 Measuring the Curie Temperature

A very important experiment is to determine the real sample temperature in the TEM with measurements performed with a already proven method. In our case, the Curie curve of our TEM sample was measured by *magneto optical Kerr effect* (MOKE). In the electron microscope, the sample was observed in Fresnel Lorentz mode and very slowly (about 0.2 K / min) cooled down while taking a CCD image every few Kelvin. Comparing both experiments delivered very promising results, one example is given in fig. 7.4. The according sample is a 50 nm thick layer of PdFe with approx. 6 at.% Fe covering the whole membrane.

In the TEM micrographs, three different micromagnetic configurations could be observed: No magnetic structures above the Curie temperature indicated by the MOKE curve, a pebbly structure just appearing at 183 K which is right at the inflection point of the MOKE measurement which is defined as T_C and the well known ripple structure at all temperatures below T_C which tells us that the film is uniformly magnetized in a direction perpendicular to the ripples.

We can learn two things from this experiment: First, the difference between the real sample temperature and the temperature reading at the controller is within $\pm 5\,\mathrm{K}$ in

---

[2]Poly(methyl methacrylate) also called poly(methyl 2-methylpropenoate)
[3]Methyl isobutyl ketone

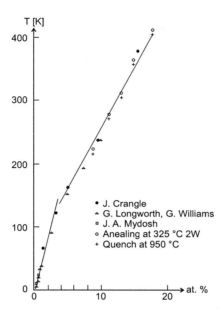

Figure 7.3: *Roundup of the dependence of the Curie temperature of PdFe of the iron concentration. For small concentrations (< 4 at.% Fe), T_C rises by 37 K per 1 at.% Fe, for higher percentages the rise is around 20 K per at.% Fe. Data from [Wu73].*

the case of the liquid nitrogen holder. The liquid helium holder could not be used for observing the phase transition of PdFe because it cannot be operated safely above 100 K. Second, even the faint magnetization of PdFe right at T_C or slightly below delivers sufficient contrast in the Fresnel mode.

7.4 Temperature Distribution on Si_3N_4 Membranes

After this general experiment with a en block film of PdFe on the Si_3N_4-membrane, the question arises if there is any spatial temperature gradient when using a pattern of small, separated magnetic structures. By observing the ferromagnetic phase transition of small particles located all over the membrane in detail, it can be demonstrated that the temperature varies less than 1 K over the entire electron transparent window of $100 \ \mu m \ \cdot \ 100 \ \mu m$.

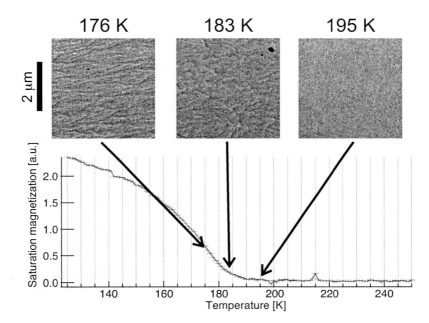

Figure 7.4: *Curie curve of a PdFe samples with ≈ 6 at.% Fe measured by MOKE. The Lorentz TEM micrographs captured at the denoted indicated temperatures show the development of a magnetic microstructure around the transition temperature. The match is very good, with no microstructure above the MOKE-T_C, first fluctuations right at T_C and a typical Ripple structure indicating an uniform in-plane magnetization below T_C.*

As a first step, a more or less quantitative indicator for the temperature has to be found. One suitable method is to use the formation of magnetic structures in *PdFe* when it changes from the paramagentic to the ferromagnetic state. The structures are observed by means of Fresnel Lorentz microscopy, which is extremely sensitive when the defocus is large enough. As the only interesting factor for this experiment is the micromagnetic structure, and therefore spatial resolution is not that important, this is no disadvantage. The experiments were done with the liquid nitrogen cooling holder, the attached temperature controller allows counter heating the sample with variable currents and a readout of the temperature of the sample holder's tip with a resolution of 1 K.

The experiment is being conducted in the following way: The sample is cooled down from well above its Curie temperature at a very slow speed so that the sample's temper-

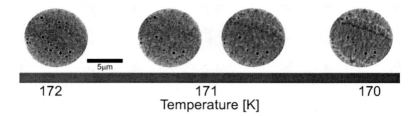

172   171   170

Temperature [K]

Figure 7.5: *Growing strength of magnetic image features of a PdFe particle when cooling down from about 172 K to about 170 K. The temperatures can not be determined exactly as the cooling stage controller only resolves 1 K steps.*

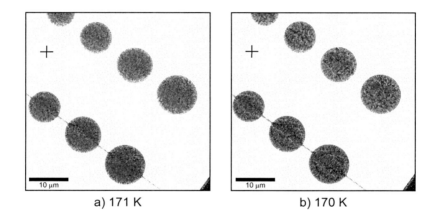

a) 171 K   b) 170 K

Figure 7.6: *Evaporated PdFe structures on a Si_3N_4-membrane around T_C. The "+" marks the center of the membrane, its rim can be seen on the lower right corner. The magnetic features are equally pronounced at the rim and in the center. The patterned connecting bars of the lower dots introduce a shape anisotropy which encourages the formation of domains.*

ature is quasi-stationary and no time-lag can be observed. By comparing the strength of the magentic image features at different indicated temperatures, the real temperatures of the observed particle can be deduced. As can be seen in figure 7.5, there are big changes in the magnetism within a temperature range of about 2 K, so that this method is good for determining the temperature of a magnetic particle from T_C to about $T_C - 2$ K at a relative uncertainity of less than ±1 K.

Figure 7.6 shows connected and not connected magnetic dots on the membrane, seated at varying locations from close to the rim to the center of the membrane. While fig. 7.6 a is taken very close to T_C, fig. 7.6 b already shows a distinct magnetic pattern. The bars connecting the lower discs favor the formation of domain walls along the bar, this is the reason why the structures inside these dots seem to be more pronounced compared to those of the non connected dots. Varying bar widths from 0.5 μm up to 5 μm have shown no influence on the temperature dependence. The magnetic structures of the dots in the center of the membrane are equally pronounced than those of the dots at its rim. This tells us that the temperature is the same within ±1 K, no matter where the observed structure is located.

7.5 Domain Formation in PdFe Squares

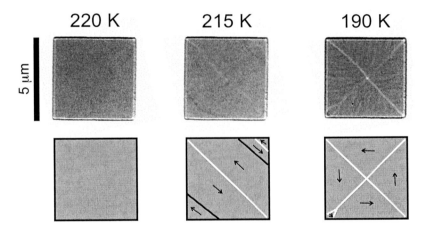

Figure 7.7: *Micrographs of a PdFe square around $T_C \approx 213\,K$. The sketches below the micrographs clarify the domain patterns and the arrows indicate the magnetization direction within a domain. Right below the Curie temperature, a pattern of five diagonal aligned stripe domains forms that transforms into the well known landau pattern at lower temperatures which minimizes the stray field energy.*

Squares of PdFe with an edge length around 5 μm show an interesting change in their domain configuration when lowering the temperature from T_C to a few Kelvin below the transition temperature.

An example is shown in fig. 7.7. The first domain configuration that forms when changing from the paramagnetic to the ferromagnetic state is a pattern of several stripe domains separated by 180°-walls. This is surprising, as such a domain pattern produces a stray field outside the sample which is at first glance not the configuration with the lowest energy. At lower temperatures, the typical Landau pattern appears, which cancels out the stray field except at the crossing of the 90°-walls where a Bloch-line is situated. But nevertheless, the stripe configuration must have the lowest total energy at this temperature.

A possible explanation for this behavior can be found when taking the fact into account that the magnetization is the parameter that changes most at T_C and a few degrees below and that the stray field is proportional to the magnetization. As the energy development with temperature cannot be calculated analytically, we compare this situation to numerical simulations in the following part.

7.5.1 Micromagnetic Simulations

The basis of micromagnetic calculations is the *Landau-Lifschitz-Gilbert* (LLG) equation which is the equation of motion for a magnetic moment $\vec{m}$ in an external field:

$$\frac{\mathrm{d}}{\mathrm{d}t}\vec{m} = -\frac{|\gamma_0|}{1+\alpha^2}\left(\vec{m}\times\vec{H}_{eff}\right) - \frac{\alpha|\gamma_0|}{1+\alpha^2}\left(\vec{m}\times\left(\vec{m}\times\vec{H}_{eff}\right)\right), \qquad (7.1)$$

with the gyromagnetic ratio γ_0 and the phenomenological damping parameter α [Höl04]. What this formula tells us is that whenever the effective field $\vec{H}_{eff}$ changes, the magnetic moment precesses around this field, while the damping makes it converge to $\vec{H}_{eff}$. The effective field can be calculated from the exchange field, the stray field and the external field (compare [Hin02]).

The simulation program LLG [Sch09a] divides the sample into equally sized cubic cells and assumes that the magnetization is homogeneous within one cell. This concretization makes only sense when the size of the cells in all directions is smaller than the magnetic exchange length l_{ex} of the simulated material. The exchange length can be calculated from the exchange constant A and the saturation magnetization M_S:

$$l_{ex} = \sqrt{\frac{2A}{\mu_0 M_S^2}}. \qquad (7.2)$$

Unfortunately, the exchange constant is not known for PdFe with a few at. % of iron, but as the magnetic iron atoms are fare away from each other, it has to be considerably smaller than the value for pure iron (which is $21\cdot10^{-12}\,J/m$) and of course >0, because PdFe is ferromagnetic.

As we do not want to simulate PdFe in detail, but rather reproduce the change of its domain configuration, the precise value of A is not so important and we use $1\cdot10^{-13}\,J/m$ as an estimation. The size and thickness of the simulated sample is the same as the real one, with an edge length of 5 μm and a thickness of 50 nm. A small magnetic bias field of 1 mT in x- and 1 mT in y-direction was applied because the existence of a small magnetic field during the experiments could not be completely ruled out.

Using these parameters, equation 7.1 is being integrated iteratively again and again until the changes of the magnetization are below a certain threshold, which is defined as the state of equilibrium. As there is no working implementation for the effects of the temperature available in the LLG software, the value of M_S was changed for each simulation and thereby used as parameter for the different temperatures.

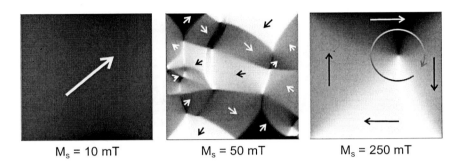

$M_s = 10$ mT $\qquad$ $M_s = 50$ mT $\qquad$ $M_s = 250$ mT

Figure 7.8: *Simulated domain structures for different saturation magnetizations (the edge length of the square is 5 μm). Different greyscales visualize different directions of the in-plane magnetization, this also is emphasized by the arrows. The left image shows a nearly uniform magnetization, while for a slightly larger M_S multiple domains appear. At a significantly larger M_S, a shifted Landau pattern develops.*

Fig. 7.8 shows the results of the simulation. The smallest saturation magnetization which corresponds to a temperature just below the Curie temperature leads to a nearly uniform magnetization distrubution which quickly changes into a multi-domain structure at a slightly higher M_S or slightly lower temperature, respectively. Increasing M_S by a factor of 5 produces a Landau pattern which is shifted because of the bias field.

Comparing the simulations to the micrographs shown in fig. 7.7, a general conformity is obvious. The deviations between simulation and measurement have many possible causes. The parameter for the exchange constant is an educated guess as its real value is not known. Furthermore, [Die08] has shown that magnetic patterns bend the Si_3N_4-membrane and thus get bent themselves resulting in a different magnetization configuration compared to flat structures. Finally, the bias field in the experiment is not known precisely and the estimated 1.4 mT in diagonal direction may be too far away from the experimental value. The visibly different position of the Bloch-line in the Landau configuration is a hint on that. More and possibly better simulations could not be performed as the available computing power is limited and each new simulation with changed parameters takes several days up to several weeks.

Nevertheless, the simulated patterns show enough resemblance to supports the thesis that for weak magnetizations close below T_C, the Landau configuration is not the energetically preferred domain pattern.

7.6 Determining the Saturation Magnetization

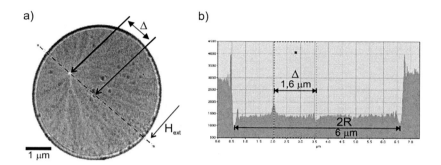

Figure 7.9: *a) Circular PdFe dot with shifted vortex due to an external field $\vec{H}_{ext}$. The arrows mark the shifted and non-shifted vortex positions. b) Line-scan along the dashed line of a). The vortex shift Δ and the dot diameter are measured from this graph.*

The saturation magnetization is one material specific parameter that can be determined with Lorentz microscopy. As described in theory chapter 3.3.1, the measurement is based on the displacement of a magnetic vortex core in respect to an external in-plane field according to formula 3.19

$$M_S = \frac{RH_\Delta}{\Delta} \left/ \left[\frac{t}{2\pi R} \left(\ln\left(\frac{8R}{t}\right) - \frac{1}{2} \right) \right] \right. .$$

This measurement was performed for a disc with a diameter of 5 μm at temperatures of 133 K and around 115 K. The external field was set to three different values. Hereby, the smallest value should deliver the most precise result, as the formula is only valid for small shifts in respect to the radius of the particle. Fig. 7.9 depicts one of the measurements. The distances are measured with the help of a line scan through the center of the magnetic disc (fig. 7.9 b). As you can see, the diameter of the disc is slightly larger than the real 5 μm, this comes mainly from the defocus but does not affect the evaluation as the distance of the vortex shift changes in the same way. Table 7.1 summarizes the experimental data of this experiment.

The results show a wide range for M_S between 225 mT and 390 mT for the higher temperatures which tells us that the approximations done during the derivation of the formula are too rough for larger vortex shifts. A second factor is that the determined

Temperature [K]	External field [mT]	Vortex shift [% of R]	Saturation magnetization [mT]
133	1.0	24	225
133	2.1	36	300
133	3.1	57	280
116	1.0	28	260
115	2.1	36	390
114	3.1	52	380
101	1.0	28	1000
101	2.1	36	194
101	3.1	52	412

Table 7.1: *Calculated saturation magnetization of a PdFe disc with a diameter of 5 μm. The error in the calculation becomes bigger with larger vortex shifts. At 101 K, pinning effects are appearing which lead to wrong data.*

saturation magnetization is very sensitive to the measured distances itself - as the uncertainty is at least 2 pixels, this alone leads to a variation of ±50 mT. At 101 K, strong effects caused by *pinning*[4] can be observed which do not play a role at the higher temperatures. The calculated saturation magnetization does not make sense in that case, as pinning is of course not considered in the formula. That is why this measurement can only tell us an estimate of the saturation magnetization in the range of 200 mT ± 100 mT. A second thing this raw measurement shows us is the tendency of M_S to become larger with decreasing temperature, as we would expect from the general form of the Curie magnetization curve.

7.7 Jumping Domain Walls

For very small magnetic particles (typically < 10 nm), the energy needed to switch their magnetization is in the range of the thermal energy $k_B T$. This statistical fluctuation of the magnetization is called *superparamagnetism*. The majority of the patterned PdFe samples with significantly larger structures has shown a behavior quite similar to superparamagnetism.

Depending on the actual temperature of the sample, some domain walls randomly jump between two (or even more) positions at dwell times of typically less then one second. This happens due to an energy landscape similar to the scheme of fig. 7.10, where the two domain wall positions a and b are local energy minima seperated by an energy barrier which is close to the thermal energy. This makes it possible for the domain wall to jump between those positions. Observing this fluctuations in the TEM is quite

---

[4]Pinning is the effect that magnetic structures get trapped at e.g. impurities or structural irregularities, compare [Die04].

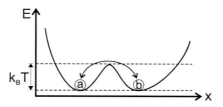

Figure 7.10: *Schema of the magnetic energy landscape which enables a domain wall to jump between the two positions a and b. The local minima are separated by an energy barrier in the range of the thermal energy $k_B T$.*

tricky as on the one hand, the flourescent screen shows a significant afterglow which leads to a smearing of the image and thus makes it impossible to differentiate between a fast movement and a statically broadened structure. The CCD on the other hand is a so called slow scan camera and only delivers micrographs with reasonable noise and contrast for shutter speeds of 0.2 s or more. The only way to do Lorentz microscopy at a sufficient frame rate is the use of the special *cinema mode* of the Gatan TRIDIEM CCD camera which delivers frames at time intervals down to 0.01 s.

Fig. 7.11 shows an example of jumping domain walls for a structure with a multi Landau domain configuration. The two possible states are depicted in a) and b). In order to make the two configurations visible, b) was subtracted from a) which leads to d), showing only the differences between the two micrographs, in our case a V-shaped structure. Image c) shows a snapshot of the jump in action and e) is a sketch of this micrograph. The temporal distribution of the positions a) and b) is shown in fig. 7.12 at a resolution of 0.02 s. This graph was acquired by the help of a custom script for the Gatan Digital Micrograph software which continuously records the intensity of a small group of pixels. This group was placed at one position of the black switching domain wall leading to a low intensity when the wall is at that position and a higher intensity when it is at the opposite one. The measurement shows us that the two positions are occupied totally random, so it makes no sense to speak of a certain hopping frequency.

This behaviour could be observed for all types of microstructures where classical domain walls or a vortex were present. Changing the temperature makes other walls jump, but for the whole temperature range below T_C achievable with the liquid nitrogen holder, the existance of jumping structures could be proven. This instability of the magnetic configuration has to be minded whenever PdFe is used as a contact for spin experiments.

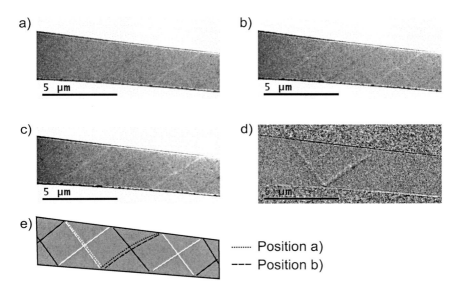

a)

5 μm

b)

5 μm

c)

5 μm

d)

5 μm

e)

········ Position a)

--- Position b)

Figure 7.11: *Micromagnetic structure with multiple Landau configuration at 160 K. a) and b)*
show the two possible configurations. c) is a snapshot with both configurations at
once, so the jump was just happening when this image was taken. d) emphasizes
the differences between a) and b) as it was generated by subtracting one from the
other. In e), the situation is sketched in order to make it easily visible.

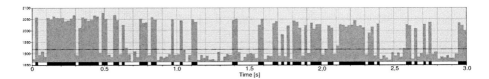

Figure 7.12: *Temporal distribution of position a (low intensity or black bar) and b (high inten-*
sity). The dashed line is the intensity value used to distinguish between the two
positions.

8 Nb$_3$Sn

Within a collaboration with G. Schierning and R. Theissmann from the University of Duisburg-Essen, the superconducting Nb3Sn ($T_C = 18$ K) was investigated. The aim of the Duisburg-Essen group is to show that a twinned lattice with domains in the nanometer range is a necessary, but yet not sufficient, condition for the occurrence of (high-T_C) superconductivity. This thesis is supported by the fact that a crystallographic phase transformation, especially of martensitic type (see explanation below), seems to be a quite universal characteristic of superconductivity. While this is a widely known fact, the microscopic structure of the forming domains is not known in detail for most superconductors.

Nb3Sn, too, performs such a martensitic phase transformation at $T_M = 43$ K, determined by a single crystal study with film technique in [Mai67], but no direct experimental evidence of twinning was given so far. Its structural change at low temperatures is investigated with the TEM in order to obtain detailed information on the defect structure of Nb3Sn.

8.1 Martensitic Transformations and Twinning

The following section provides a brief introduction to martensitic transformations, details can be found in [Bha03]. The martensitic transformation is a solid to solid phase transformation where the lattice or molecular structure changes abruptly at some temperature.

Fig. 8.1 a) shows the high temperature state of a sample lattice, where the atoms are arranged in a square. When the temperature is lowered more and more, at first the solid will show the usual thermal contraction. But at a certain critical temperature, the crystal lattice structure changes abruptly e.g. to the rectangular structure shown in fig. 8.1 b). Even though this change is abrupt and the unit cell is quite significantly distorted, the next neighbourhood of each atom stays the same, so there is no diffusion. This transformation is called the martensitic phase transformation; it is displacive and of first order. The high temperature phase is called the *austenit* phase, while the low temperature phase is called the *martensite* phase. If the temperature is lowered even more, nothing irregular happens, only the usual thermal contraction again. When heating up the crystal again, the opposite effect can be observed: ordinary thermal expansion until a certain critical temperature is reached and the lattice snaps to its high temperature phase again. The transition temperatures for the heating or cooling process usually differ by a few to several hundred degrees, depending on the material. Another characteristic feature of the martensitic transition is the formation of microstructures. Usually, the austenite phase shows a higher symmetry than the corresponding martensite phase. This leads to multiple symmetry-related variants of martensite as shown in 8.1, where the square lattice can transform either one of the rectangular lattices b) and c). The change of symmetry during the transformation determines the number of variants.

In real life, the crystal transforms into a mixture of different martensite variants because of independent nucleation events and energy minimization. This mixture has to be done in a coherent way, this means that the rows of atoms have to stay unbroken across the interfaces of the variants. Fig. 8.1 d) shows such an arrangement of two rectangular lattices arranged in a manner that the rows of atoms are kinked but unbroken. The resulting microstructure is characteristic for the material, its size, the grain size and the history of the sample. The length-scale of the structures ranges from a few nanometers to tenths of millimeters.

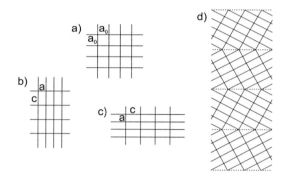

Figure 8.1: *Schematic illustration of the martensitic phase transformation. a) shows the austenite physe, b) and c) different martensite variants and c) a possible arrangement of alternating martensite variants. After [Bha03].*

8.2 Sample Preparation

The sample is a polycrystalline Nb_3Sn pellet prepared from commercially available powder by hot pressing. For the TEM study, the pellet was prepared by grinding, polishing and Ar-ion milling as already described in chapter 6.2.1.

8.3 Pre Characterization

In order to be able to interpret the subsequent low temperature TEM investigations, the Nb_3Sn sample was pre characterized in detail.

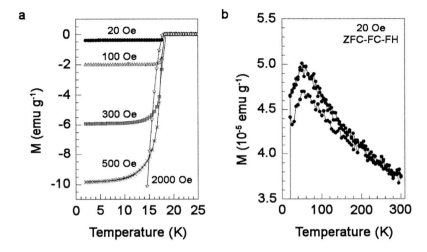

Figure 8.2: *a: Meissner-Ochsenfeld effect of the Nb_3SN sample. The transition temperature is around 18 K.*
b: Zero field cooled, field cooled, field heated series of measurements with B = 20 Oe. The maximum at 50 K indicates less paramagnetic or more diamagnetic contributions to the net susceptibility below 50 K. Image from [Sch09b].

Low temperature neutron diffraction on this sample performed at the FRM2 reactor in Garching confirmed that there is a phase transition between 60 K and 30 K. Magnetic measurements by a SQUID [1] on the one hand reproduced the literature value of $T_C = 18$ K, but on the other hand showed a maximum in the magnetization curve at about 50 K (see fig. 8.2). This behavior can be interpreted as a change in the magnetic susceptibility by a crystallographic phase transition.

8.4 TEM Study

The sample showed a structural change between 26 K and 14 K, visible by the formation of stripe contrast. This contrast appeared when the sample was cooled below 22 K and vanished at higher temperatures. The stripes are interpreted as structural domains separated by domain walls. The width of the domains is about 30 nm. As one can see in fig. 8.3, the appearance and vanishing of the stripes could be reproduced several times. The transition temperature of 22 K differs from the temperatures determined by

---

[1] *Superconducting QUantum Interference Device*

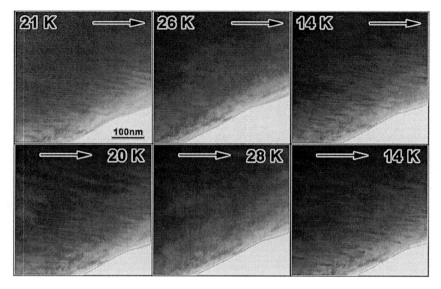

Figure 8.3: *Temperature series of Nb_3Sn in the TEM. Stripe-contrast is visible below 22 K. The formation of a similar stripe contrast is perfectly reversible upon cooling and heating cycles.*

neutron diffraction and SQUID, but this can be explained satisfactorily: First of all, the temperature reading of the liquid Helium holder has an uncertainty of at least 10 K (compare chapter 5.2). Second, the TEM sample is observed in an area with a remaining thickness of < 500 nm, compared to the thick bulk samples investigated by the other methods. The thin sample has a much lower accumulated strain within the material which leads to a lower transition temperature.

The transmission electron micrographs show indications that nanodomains formed by a crystallographic phase transformation are a feature of the low temperature real structure of Nb_3Sn. This finding supports the thesis of the Duisburg-Essen group and could be a first step towards a better understanding of the origin of superconductivity.

9 Conclusions and Outlook

Within this work, magnetic and structural phase transitions at low temperatures could be investigated. Thereby, the new Gatan liquid helium cooling holder was introduced and its performance and limits in combination with our Tecnai F30 microscope were evaluated. Its faint but steady vibrations made it impossible to use it for electron holography, which is a big drawback. This vibrations are a characteristic for the older liquid nitrogen cooled holder, too. The Tecnai microscope with its current compustage and the mechanically loose sample holder clamping has to be changed fundamentally in order to allow electron holography at low temperatures. The problems arising from the fact that the liquid helium holder always has a different angle in respect to the compustage because of its heavy wight, could be solved with the newly developed α-tilt measurement device.

The investigated materials were gallium manganese arsenide (Ga,Mn)As and palladium iron (PdFe) for the ferromagnetic transition and niobium-3-tin (Nb3Sn) for the crystallographic one.

The difficult sample preparation of (Ga,Mn)As taught us new preparation methods, amongst them the splinter technique and the very promising chemical etching procedure. This material also triggered the investigations of the temperatures arising during the sample preparation by ion milling. The new low-energy guns in addition with a liquid nitrogen cooling unit our ion mill was recently equipped with should reduce thermal stress on future samples. Furtheron, a new diploma thesis still in progress concerning the temperature rise of a TEM sample in the microscope during electron iradiation by various methods like lithographic thermo couples and bimetallic microstructures.

This topic is very interesting, because it could help in finding a conclusive answer why we could not find any magnetic features of the (Ga,Mn)As samples with Lorentz TEM. Maybe future samples with a significantly higher Curie temperature will lead to success. The difficulties in observing the magnetism of (Ga,Mn)As in the TEM led to a new method of electron holography which eliminates all incidents and may be used for any kind of holographic stray field imaging in the future.

PdFe has Curie temperatures which are high enough to be reached with the liquid nitrogen cooled sample holder. Unlike to the material before, imaging magnetic features was successful with PdFe. It allowed us to measure the Curie temperature of the samples and compare it with MOKE measurements. This showed us that in the case of the liquid nitrogen holder and PdFe samples grown on Si_3N_4-membranes, the indicated temperature on the holder controller and the sample temperature match within less than $\pm$ 5 Kelvin. By structuring several small PdFe particles all around the membrane's surface, we could measure the lateral temperature distribution over the membrane. We have found that in this special case, there is no lateral temperature gradient within the accuracy of our experiment of ± 1 Kelvin.

While watching the phase transition in detail, surprising domain patterns appeared just below T_C for square particles of a few μm edge size. Micromagnetic simulations with LLG showed that they originate from the rivalry between mainly the exchange energy and the stray field energy which is direct proportional to the saturation magnetization. As M_S is the lower the closer the temperature is to T_C, there is no need for the magnetic configuration to form closure domains. The sequence of the domain configuration, coming from high temperatures, is paramagnetic (above T_C), single domain, stripe domains and finally the well known Landau pattern.

Another effect we could detect with most samples is that some domain walls are randomly jumping between two positions – which walls are affected depends on the temperature of the sample. The reason for this behavior is that the magnetic energy landscape depends on the temperature, and for a certain temperature and location, the two positions of the domain wall are separated by an energy barrier which has a height closely to the thermal energy. The hopping of the walls thus is thermally excited and has a great similarity with superparamagnetism.

The measurement of the shift of the magnetic vortex due to an external field allowed us to determine the saturation magnetization for different temperatures. As the calculation of the value is extremely sensitive on the measured shift distance and only valid for shifts small compared to the radius of the disc, this method only delivered a very rough estimation of M_S.

Within a collaboration with the University of Duisburg-Essen, we validated the existence of a martensitic phase transition of Nb$_3$Sn with twinning. They want to prove that the existence of such a transition is a necessary requirement for the development of superconductivity. The microscopic studies successfully showed such a crystallographic transition at temperatures around 22 K, which is close to but not precisely the expected temperature. The reason for the difference most probably is a shifted transition temperature because of the very thin TEM sample and a difference between the real sample temperature and the temperature indicated by the liquid helium holder controlling unit.

In the case that the thesis of the Duisburg-Essen group solidifys, the martensitic phase transition of many superconducting materials has to be investigated in detail. This makes it necessary to use a sample holder that can reach even lower temperatures than the current one with 12 K minimum temperature. Achieving temperatures of a few Kelvin at the illuminated specimen area in the TEM seems to be still difficult and requires a lot of developement for both holders and microscopes.

Bibliography

[Ade05] M. ADELL, J. KANSKI, L. ILVER ET AL. Comment on "Mn Interstitial Diffusion in (Ga,Mn)As". *Physical Review Letters*, 94:139701 (2005).

[Bha03] K. BHATTACHARYA. Microstructure of Martensite. *Oxford University Press* (2003).

[Bro62] W. F. BROWN. Magnetostatic Principles in Ferromagnetism. *North Holland Publishing Company* (1962).

[Cha84] J. N. CHAPMAN. The investigation of magnetic domain structures in thin foils by electron microscopy. *Journal of Physics D*, 17:623 (1984).

[Die04] C. DIETRICH. Lorentzmikroskopische Untersuchungen zu Lage und Pinning magnetischer Vortices. *Diplomarbeit Universität Regensburg* (2004).

[Die08] C. DIETRICH, R. HERTEL, M. HUBER ET AL. Influence of perpendicular magnetic fields on the domain structure of permalloy microstructures grown on thin membranes. *Phys. Rev. B*, 77(17):8 (2008).

[Edm04] K. EDMONDS, P. BOGUSŁAWSKI, K. WANG ET AL. Mn Interstitial Diffusion in (Ga, Mn) As. *Physical Review Letters*, 92(3):37201 (2004).

[Geb80] GEBHARDT AND KREY. Phasenübergänge und kritische Phänomene. *Vieweg & Sohn Verlagsgesellschaft mbH* (1980).

[Heu05] M. HEUMANN. Elektronenholografie an magnetischen Nanostrukturen. *Dissertation Universität Regensburg* (2005).

[Hin02] D. HINTZKE. Computersimulationen zur Dynamik magnetischer Nanostrukturen. *Dissertation Gerhard-Mercator-Universitaet Duisburg* (2002).

[Hof65] H. HOFFMANN. Mikromagnetische Theorie der quasistationären Eigenschaften dünner Schichten. *Habilitationsschrift München* (1965).

[Höl04] R. HÖLLINGER. Statische und dynamische Eigenschaften von ferromagnetischen Nano-Teilchen. *Dissertation Universität Regensburg* (2004).

[Hur04] C. HURM. Bestimmung des magnetischen Detektionslimits in der Lorentzmikroskopie. *Diplomarbeit Universität Regensburg* (2004).

[Hur08]   C. HURM. Towards an unambiguous Electron Magnetic Chiral Dichroism (EMCD) measurement in a Transmission Electron Microscope (TEM). *Dissertation Universität Regensburg* (2008).

[Kai00]   S. KAISER. TEM-Untersuchungen von heteroepitaktischen Gruppe III- Nitriden. *Dissertation Universität Regensburg* (2000).

[Kas56]   T. KASUYA. *Prog. Theor. Phys.*, 45 (1956).

[Kit02]   C. KITTEL. Einführung in die Festkörperphysik. *Oldenburg Verlag München Wien* (2002).

[Lan06]   R. LANG. Magnetooptische Untersuchungen an ferromagnetischen GaMnAs-Epitaxieschichten. *Dissertation Universität Bayreuth* (2006).

[Mai67]   R. MAILFERT, B. W. BATTERMAN AND J. J. HANAK. Low temperature structural transformation in Nb3Sn. *Phys. Lett.*, 24A(6):315 (1967).

[Ohn96]   H. OHNO, A. SHEN, F. MATSUKURA ET AL. *Appl. Phys. Lett.*, 69:363 (1996).

[Ohn99]   H. OHNO. *J. Magn. Mater.*, 200:110 (1999).

[Ott01]   S. OTTO. Konstruktion, Bau und Charakterisierung eines Magnet-Probenhalters für das Elektronen-Mikroskop. *Diplomarbeit Universität Regensburg* (2001).

[Pot01]   S. POTASHNIK, K. KU, S. CHUN ET AL. Effects of annealing time on defect-controlled ferromagnetism in GaMnAs. *Appl. Phys. Lett.*, 79:1495 (2001).

[Rae05]   H. RAEBIGER, A. AYUELA AND J. VON BOEHM. Electronic and magnetic properties of substitutional Mn clusters in (Ga, Mn) As. *Phys. Rev. B*, 72(1):14465 (2005).

[Rei97]   L. REIMER. Transmission Electron Microscopy. *Springer Verlag Heidelberg* (1997).

[Rud54]   M. A. RUDERMAN AND C. KITTEL. *Phys. Rev.*, 96:99 (1954).

[Sad06a] J. SADOWSKI, J. Z. DOMAGAŁA, J. KANSKI ET AL. How to make GaMnAs with a high ferromagnetic phase transition temperature? *Materials Science-Poland*, 24(3):9 (2006).

[Sad06b] J. SADOWSKI, J. Z. DOMAGAŁA, V. OSINNIY ET AL. High ferromagnetic phase transition temperatures in GaMnAs layers annealed under arsenic capping. *arXiv:cond-mat/0601623v2* (2006).

[Sap02]   V. SAPEGA, M. MORENO, M. RAMSTEINER ET AL. *Phys. Rev. B*, 66 (2002).

[Sch03]   M. SCHNEIDER. Untersuchungen des Ummagnetisierungsverhaltens von polykristallinen Mikro- und Nanostrukturen aus Permalloy mit der Lorentz-Transmissionselektronenmikroskopie. *Dissertation Universität Regensburg* (2003).

[Sch09a]   M. R. SCHEINFEIN. LLG Micromagnetics Simulator. *http://llgmicro.home.mindspring.com* (2009).

[Sch09b]   G. SCHIERNING, R. THEISSMANN, J. GRÜNDMAYER ET AL. Low temperature transmission electron microscopy study of superconducting Nb3Sn. *not yet published* (2009).

[Smi70]   T. F. SMITH, W. E. GARDNER AND H. MONTGOMERY. A study of spin-wave excitations in dilute Pd-Fe alloys. *J. Phys. C: Metal Phys.*, 3 (1970).

[Sta05]   V. STANCIU, O. WILHELMSSON, U. BEXELL ET AL. Influence of annealing parameters on the ferromagnetic properties of optimally passivated (Ga, Mn) As epilayers. *Phys. Rev. B*, 72(12):125324 (2005).

[Sug08]   A. SUGAWARA, H. KASAI, A. TONOMURA ET AL. Domain Walls in the (Ga,Mn)As Diluted Magnetic Semiconductor. *Phys. Rev. Lett.*, 100(4):4 (2008).

[Twa99]   A. TWARDOWSKI. *Mat. Sci. Eng. B*, 63:96 (1999).

[Wan04]   K. WANG, R. CAMPION, K. EDMONDS ET AL. Magnetism in (Ga,Mn)As Thin Films With TC Up To 173K. *cond-mat/0411475* (2004).

[Wen08]   M. WENGBAUER. Bestimmung der Temperaturbelastung von TEM-Proben bei der Präparation mit Ionenstrahlen. *Diplomarbeit Universität Regensburg* (2008).

[Woj04]   T. WOJTOWICZ, J. FURDYNA, X. LIU ET AL. *Physica E*, 25:171 (2004).

[Wu73]   Y.-C. WU, J. CHEN AND S.-H. FANG. Determination of Curie Temperature, Short Range Order Parameter, and Number of Nearest Neighbors of Palladium Atoms to Iron Atoms in Palladium-Iron Alloys by Resistivity Measurement. *Chinese Journal of Physics*, 11(2):370 (1973).

[Xu05]   J. XU, M. VAN SCHILFGAARDE AND G. SAMOLYUK. Role of Disorder in Mn: GaAs, Cr: GaAs, and Cr: GaN. *Physical Review Letters*, 94(9):97201 (2005).

[Yeh75]   B.-H. YEH, J. CHEN, P. K. TSENG ET AL. MagnetizationMeasurements of Pd-Fe Alloys in the Intermediate Fe Concentration Region. *Chinese Journal of Physics*, 13(1):1 (1975).

[Yos57]   K. YOSIDA. *Phys. Rev.*, 106:893 (1957).

[Yu02]   K. M. YU, W. WALUKIEWICZ, T. WOJTOWICZ ET AL. Effect of the location
          of Mn sites in ferromagnetic GaMnAs on its Curie temperature. *Physicl Review
          B*, 65:201303(R) (2002).

[Zwe92]   J. ZWECK. Moderne elektronenmikroskopische Untersuchungsmethoden in der
          Festkörperphysik. *Habilitationsschrift Universität Regensburg* (1992).

A Acknowledgment

Ein herzliches Dankschön an alle, die mich bei dieser Arbeit unterstützt haben - und ganz besonders an...

... Joe Zweck für das spannende Thema, die Tagungen im In- und Ausland, die vielen Spezialaufträge, die lustigen Geschichten ...

... Christian Hurm, Christian Dietrich, Martin Beer und Marcello Soda für die guten Unterhaltungen und den Spaß, den wir immer hatten.

... meine Zimmerkollegen Stephanie Hussnätter, Christoph Kefes, Roman Grothausmann, Martin Wengbauer, Clemens Sill und Matthias Lohr für die angenehme Atmosphäre.

... meinen ehemaligen Innovator-Schüler Martin Heumann, der seinen Meister zu wahren Höchstleistungen gebracht hat.

... alle anderen aktiven und ehemaligen Mitglieder der TEM-Gruppe, Tom Uhlig, Martin Brunner, Thomas Haug, Michael Binder, Karl Engl, Florian Herzog, Sebastian Liebl, Pavel Kasanzky, Johannes Thalmair, Michael Huber, Christian Huber, Andreas Hasenkopf, Martin Müller, Josef Lang, Jörg Köhler, Florian Schneider, Rudolf Simml, Christoph Hansch und ganz besonders an den "guten Geist" im Labor, Olga Ganitcheva.

... Korbinian Perzlmaier für die Unterstützung bei theoretischen Fragen.

... Ursula Wurstbauer, Martin Utz, Ondrej Vavra, Janusz Sadowski und Eva Brinkmeier für die vielen vielen Proben.

... Christian Dietrich für die viele Arbeit im Reinraum.

... Matthias Sperl, Mathias Kiessling und Marcello Soda für die vielen Messungen.

... Gabi Schierning und Ralf Theissmann von der Uni Duisburg-Essen für die gute Zusammenarbeit.

... die Mitarbeiter der Mechanikwerkstatt, die auch die kniffligsten Aufträge meisterten, ganz besonders Walter Wendt.

... die Mitarbeiter der Elektronikwerkstatt, auch für die Erlaubnis, ihre Geräte mitzubenutzen.

... Herrn Harms und den anderen Service-Technikern der Firma FEI, ohne die schon lange kein Strahl mehr eine Probe treffen würde.

... Christa Mayer für ihre Hilfe und die gute Schokolade.

... den gesamten Lehrstuhl Back für die gute Zusammenarbeit.

... alle Angehörigen des SFB 689 für die vielen informativen Vorträge und Gespräche sowie ihre freundlichen Hilfestellungen.

... die Lehrstühle Weiss und Wegscheider für die Möglichkeit, ihre Einrichtungen zu nutzen.

... unsere Kolonialwarenhändler Dieter Schierl und Tobias Stöckl.

... Hans-Georg Pankau und Bernd Kraus von der Firma Gatan für ihre Hilfsbereitschaft bei Problemen mit dem Probenhalter und der Programmierung von Digital Micrograph.

... Christian Hurm und Simon Wachter fürs Korrekturlesen.

... meine Eltern, die mich während der gesamten Zeit immer unterstützt haben.

Ganz besonders möchte ich meiner Kathi für ihre Liebe und Geduld danken.

Bisher erschienene Bände der Reihe
„Applied Electron Microscopy - Angewandte Elektronenmikroskopie"

ISSN 1860-0034

Alle erschienenen Bücher können unter der angegebenen ISBN direkt online (http://www.logos-verlag.de/Buchreihen) oder per Fax (030 - 42 85 10 92) beim Logos Verlag Berlin bestellt werden.